Advances in Electronic Marketing

Irvine Clarke III
James Madison University, USA

Theresa B. Flaherty
James Madison University, USA

IDEA GROUP PUBLISHING

Hershey • London • Melbourne • Singapore

Acquisitions Editor:	Renée Davies
Development Editor:	Kristin Roth
Senior Managing Editor:	Amanda Appicello
Managing Editor:	Jennifer Neidig
Copy Editor:	April Schmidt
Typesetter:	Marko Primorac
Cover Design:	Lisa Tosheff
Printed at:	Integrated Book Technology

Published in the United States of America by

MK

    Idea Group Publishing (an imprint of Idea Group Inc.)
    701 E. Chocolate Avenue, Suite 200
    Hershey PA 17033
    Tel: 717-533-8845
    Fax: 717-533-8661
    E-mail: cust@idea-group.com
    Web site: http://www.idea-group.com

and in the United Kingdom by

    Idea Group Publishing (an imprint of Idea Group Inc.)
    3 Henrietta Street
    Covent Garden
    London WC2E 8LU
    Tel: 44 20 7240 0856
    Fax: 44 20 7379 3313
    Web site: http://www.eurospan.co.uk

    Library of Congress Cataloging-in-Publication Data

Advances in electronic marketing / Irvine Clarke, III and Theresa Flaherty, editors.
    p. cm.
 Summary: "This book addresses Internet marketing and the World Wide Web, and other electronic marketing tools such as geographic information systems, database marketing, and mobile advertising"--Provided by publisher.
 Includes bibliographical references and index.
 ISBN 1-59140-321-9 (h/c) -- ISBN 1-59140-322-7 (s/c) -- ISBN 1-59140-323-5 (ebook)
 1. Internet marketing. 2. Internet advertising. 3. World Wide Web. I. Title: Electronic marketing. II. Clarke, Irvine, 1961- III. Flaherty, Theresa, 1968-
 HF5415.1265.A38 2005
 658.8'72--dc22

                             2004029847

British Cataloguing in Publication Data
A Cataloguing in Publication record for this book is available from the British Library.

Advances in Electronic Marketing

Table of Contents

Section I: Buyer Behavior of Online Consumers

Eileen Bridges, Kent State University, USA
Ronald E. Goldsmith, Florida State University, USA
Charles F. Hofacker, Florida State University, USA

Alvin Y.C. Yeo, University of Western Australia, Australia
Michael K.M. Chiam, University of Western Australia, Australia

Francesca Dall'Olmo Riley, Kingston University, UK
Daniele Scarpi, Universitá di Bologna, Italy
Angelo Manaresi, Universitá di Bologna, Italy

Preface

Introduction to Advances in Electronic Marketing

While the Internet has been in existence since the 1960s, it was not until the advent of the World Wide Web (WWW) that businesses began to modify practices in an attempt to exploit the advantages of this new technology. Most early entrants primarily focused on informational activities while discovering limiting technologies and perplexing consumer behavior. Few companies were able to utilize the Internet's full capabilities, and stories of failure amassed. The transition was seldom smoothe, and companies enjoyed a mixture of successes and failures. In this period, one lesson became clear: electronic businesses operated in a distinct and unique manner from traditional business activities.

Today, the Internet has exploded into mainstream society. The electronic medium has created an informational and communication revolution that has forever changed the overall business environment. Consequently, marketers continue to explore the possibilities of electronic marketing as an ideal channel for communication, entertaining, selling, and distributing goods, services, and ideas. Marketers seek opportunities to tap into the Internet's potential for optimizing performance and success, while realizing that it remains a dynamic ever-changing medium. With a brisk rate of change, the state of electronic marketing remains fluid, success factors transform, and marketers must continually pursue the most robust advances in the field.

In the early 1990s, electronic marketing was fueled by the potential of revolutionizing the manner in which organizations conducted their entire business operations. In the past decade, with its rapid explosion of Web technologies, marketers have experienced excitement and panic; trial and error; and success and failure as they created online businesses and expanded into electronic channels of distribution. Most now agree that Internet marketing remains unique from traditional approaches, yet, still draws on fundamental marketing principles. Therefore, the genesis of this book lies in investigating contemporary marketing thought about how the Internet has changed the face of marketing. Specific emphasis is placed on managerially relevant discussions of progress in electronic marketing.

The buyer behavior of online consumers is the starting point for *Advances in Electronic Marketing*. In Section I, issues are investigated, such as customer attraction and retention, e-customer loyalty, factors influencing purchase behavior on the Internet, and customer willingness to provide information online. Section II explores emerging strategic issues associated with e-marketing. Technological tools (e.g., databases, wireless devices, and geographical information systems), once unavailable to traditional marketers but now commonplace for e-marketers, are investigated in Section III. Section IV includes two complementary chapters devoted to the various opportunities and challenges of the law as it pertains to e-marketing. Theoretical frameworks and models of e-marketing phenomena are presented in Section V.

Buyer Behavior of Online Consumers

All marketing efforts are ultimately directed toward the consumer. As such, the introductory section takes a closer look at some of the emerging issues pertaining to buyer behavior of online consumers.

Chapter I. Eileen Bridges, Ronald E. Goldsmith, and Charles F. Hofacker differentiate between business-to-business (B2B) and business-to-consumer (B2C) marketplaces. The authors describe how and why customers purchase online and why consumers are attracted to particular suppliers. Online and off-line customers are compared in order to understand reasons for observed differences. Various antecedents of the online experience are addressed to determine influences on satisfaction and buying behavior. Web site efficacy (usefulness and ease-of-use) is addressed in light of its importance in customer satisfaction and retention for online shopping. The chapter concludes with insights for e-marketers that wish to attract new buyers, satisfy, and retain them.

Chapter II. E-customer loyalty is defined as the intention to repurchase a certain product/service consistently from a particular e-vendor, despite the presence of other circumstances that may induce switching behavior. Alvin Y.C. Yeo and Michael K.M. Chiam provide an integrated framework for understanding the impact of corporate image, customer trust, and customer value on e-customer loyalty in a B2C e-commerce context. This framework incorporates cognitive, affective, and conative components in order to gain customer mind share, nurture emotional ties, and influence future purchase decisions. The authors offer managerial suggestions for online loyalty management to attain the "*tao of loyalty*".

Chapter III. This chapter provides a thorough analysis of three key factors that can influence consumer purchase behavior on the Internet. These factors include product-related factors (e.g., product type, brand name, etc.), consumer-related factors (e.g., consumer expertise, attitudes, risk perceptions, shopping orientation, etc.), and retail-related factors (e.g., strategies and tactics). Francesca Dall'Olmo Riley, Daniele Scarpi, and Angelo Manaresi propose that consumer-related factors affect online purchasing and the resulting implications for online retailers. The authors provide practical suggestions for retailers to reduce or overcome some of the barriers that prevent consumers from increasing the amount of products purchased online.

Chapter IV. The Internet has emerged as a powerful electronic customer relationship management tool. However, this tool is of practical use only when consumers are willing to provide the type of information that is of value to the e-marketer. Consumer willingness to provide personal information is a cornerstone of customer relationship management. Terry Daugherty, Matthew Eastin, and Harsha Gangadharbatla explore how consumers' self-confidence in using the Internet impacts their willingness to provide personal information online. Results from their consumer panel support the idea that increasing Internet confidence may result in more favorable attitudes toward information requests and an increased willingness to provide information.

E-Marketing Strategy

We include several unique aspects of marketing strategy in this section to highlight some of the distinctive characteristics of the Internet. The first chapter in this section accentuates the international nature of the Internet. A chapter on branding emphasizes challenges of creating interactive brands in the online and off-line worlds. The use of surprise is incorporated as an important element of viral marketing strategies in the next chapter. Finally, the mobile advertising chapter highlights the array of marketing opportunities available for use in a wireless environment.

Chapter V. The Internet and WWW provide new avenues for developing international e-marketing strategies. Wider market access, efficient resource access, global niches, competitive advantage, efficient coordination, online distribution, and production/sales smoothing are among the many strategic and operational opportunities provided by the Internet. However, as Gopalkrishnan R. Iyer discusses, several barriers exist which limit the effectiveness of global Internet marketing strategies. Challenges, such as infrastructure, laws, regulations, and culture, limit the global reach of organizations. This chapter provides insight for utilizing the Internet to deploy international marketing strategies. A managerial framework is presented to assist in the creation of customer-centric global organizations.

Chapter VI. Mary Lou Roberts makes a case for the importance of branding efforts by reviewing major approaches to brand development in both off-line and online marketing environments. The concept of Interactive Brand Experiences (IBE) is created and explored via the use of marketing tools, such as personalization, co-creation, purchase-process streamlining, self-service, brand community, rich media, product self-design, dynamic pricing, and customization. Roberts concludes that there are two major challenges involved in integrating branding efforts in online and off-line spaces: (1) identifying the appropriate techniques and the media best suited to deliver them and (2) executing seamlessly at all touchpoints in the process.

Chapter VII. Viral marketing strategies encourage customers to pass along marketing messages utilizing their own network of friends, relatives, colleagues, and so forth. Adam Lindgreen and Joëlle Vanhamme explain the primary mechanisms of viral marketing and investigate the emotion of surprise within viral marketing campaigns. The authors posit that drivers of viral marketing often include elements of surprise. Their exploratory research revealed that viral marketing campaigns should be based on surprising messages that benefit the recipients of the messages. The chapter includes a summary of major steps that marketers can implement to develop successful viral campaigns.

Chapter VIII. Mobile advertising (m-advertising) consists of marketing messages sent to and received on mobile devices, such as cell phones, personal digital assistants, and other handheld devices. Jari Salo and Jaana Tähtinen discuss unique features of m-advertising and investigate the use of permission-based m-advertising in a Finnish retail context. Their case data was derived from the SmartRotuaari service system which includes a wireless multi-access network, middleware for service positioning, a Web portal, and m-advertising services. Their research concludes that retailers and advertising agencies have much to learn about m-advertising. Future trends and suggestions are offered for effectively utilizing the potential of m-advertising.

Technology for E-Marketing

The Internet, as a marketing tool, has received much attention in both practitioner and research communities. However, little attention has been given to emerging technologies as they pertain to e-marketing. This section of *Advances in Electronic Marketing* introduces contemporary research about the use of database integration, wireless devices, and geographic information systems in current Internet marketing practices.

Chapter IX. The Internet allows marketers to collect consumer data in a fast, inexpensive, and relatively accurate manner. Sally Rao and Chris O'Leary argue that Internet/database marketing provides solutions for some of the difficulties associated with customer relationship management. Using an action research methodology, the authors develop a theoretical framework for integrating the Internet with database marketing strategy. Data is collected from the Internet and then integrated into organizational databases. This integration process involves data content identification/understanding, data integration/aggregation, and data warehousing. Internet and database marketing strategies are derived from the integration and improved through one-to-one interactivity and customization.

Chapter X. The mobile channel and how mobile phones can be utilized for mobile branding (m-branding) is explored by Jari Helenius and Veronica Liljander. While m-branding is in its infancy, it is an emerging phenomenon in current brand management. The authors propose four mobile branding techniques that can be used to strengthen brand assets: sponsored content, mobile customer relationship marketing, different forms of mobile advertising, and a mobile portal. Each of these techniques is associated with specific brand asset objectives (e.g., awareness, loyalty, associations, etc.) as well as suggestions for supportive media that can be utilized in combination with m-branding.

Chapter XI. As a relatively new technology, Mark R. Leipnik and Sanjay S. Mehta explain how Geographic Information Systems (GIS) can be applied to electronic marketing efforts. While GIS technologies have been used for almost 40 years, marketing applications, and particularly e-marketing applications, are just beginning to emerge. The chapter's objective is to help marketing managers gain an appreciation for GIS technologies for use in strategic decision making. The authors list sources of software and data to assist in developing GIS structures. Specific e-marketing examples of GIS applications are examined in industries, such as tourism, real estate, and market research.

E-Marketing Legal Challenges

The legal/regulatory environment cannot be ignored when developing and implementing e-marketing strategy. This environment is complex, changing, and in many instances, unresolved. While great efforts have been made to provide a legal environment with clear uniform law, online marketers must be aware that both domestic and global rules can form the basis for liability.

Chapter XII. As Michael T. Zugelder points out, there are numerous legal issues facing online marketers. Current law, both domestic and global, is varied and poses great potential for liability. E-marketers must contend with intellectual property concerns, such as trademark law, copyright law, guarding trade secrets, and protection of patents. Information management becomes a concern as there are many laws that regulate, control, and sometimes forbid the collection or dissemination of information. Consequently, these laws impact online advertising, consumer privacy concerns, and data collection. Zugelder's chapter concludes with a discussion of online contract formation and recent developments with contract law.

Chapter XIII. Heiko deB. Wijnholds and Michael W. Little identify and categorize major regulatory and marketing challenges as they pertain to global online marketing. The chapter emphasizes international transactions between the U.S. and European Union (EU) by comparing and contrasting marketing practices and specific regulatory positions affecting e-commerce trade. The authors highlight the U.S. and EU positions regarding regulation versus self-regulation. Policies are reviewed, such as those affecting sales taxes on Internet purchases, income taxes, liability, and jurisdiction. The chapter lists business and legal precautions to enhance e-marketing efforts and meet e-marketing objectives.

E-Consumer Theoretical Frameworks

Advances in Electronic Marketing concludes with theoretical frameworks that can be used to explore various e-marketing phenomena. Building on some of the ideas presented in prior chapters, the final section provides insight on attitudes toward advertising on the Internet, virtual communities, and online purchase decision cycles.

Chapter XIV. It is well-known in the marketing literature that consumer attitudes toward products and advertisements can impact consumer behavior. However, marketing research has largely explored this concept via traditional (off-line) means of advertising. This chapter provides an initial exploration of attitude toward advertising as applicable to advertising on the Internet. Chris Manolis, Nicole Averill, and Charles M. Brooks use experimentation to manipulate consumer attitudes toward Internet advertising and measure resulting effects on brand attitudes and intentions to purchase. Their study found that Web site advertising does not appear to significantly impact attitudes toward the brand; however, attitudes toward the brand affected purchase intentions.

Chapter XV. This chapter explains the unique features of virtual communities with special emphasis on the strategies, managerial suggestions, and technologies employed for this type of business model. Carlos Flavián and Miguel Guinalíu analyze virtual communities from a sociological perspective with particular attention to communities developed around a brand. Explanation is provided for why individuals become members of such communities, including dealing with matters of interest, establishing relationships, living out fantasies, and carrying out transactions. The authors outline recommendations for the marketing of virtual communities in order to incorporate feasible strategies and common technologies (e.g., discussion forums, e-mail groups, chat rooms, MUDs, content management, peer-to-peer systems, and really simple syndication).

Chapter XVI. Penelope Markellou, Maria Rigou, and Spiros Sirmakessis present a framework depicting an online consumer purchase decision cycle. The framework assumes that online consumers are triggered by a stimulus to purchase certain items and is based on three distinct and important phases: building trust and confidence, online purchase experience, and after-purchase needs. Their framework is used to examine customer states and transition conditions, with special focus on abandonment factors, as online consumers interact outside of the e-shop, inside the e-shop, and after the purchase. The authors provide guidance for managers by outlining success and failure factors during the online consumer purchase decision cycle.

Conclusion

Due to the fluid nature of the field, absolute answers to electronic marketing problems are unattainable. However, this book does provide a comprehensive collection of cutting-edge research on Internet and technological applications for marketing. By investigating major elements of Internet marketing, including online marketing strategy, marketing research, database development, online consumer behavior, customer relationship marketing, and the online marketing mix, readers should have a better understanding of the current state of the discipline. The primary contribution lies in bringing together a global perspective, from many of the leading researchers, of the issues facing electronic marketing today. We hope that *Advances in Electronic Marketing* will serve as a useful resource for greater understanding of the concepts, theories, practices, and current state of electronic marketing.

Acknowledgments

We wish to thank the staff of Idea Group Publishing, most notably Mehdi Khosrow-Pour, Jan Travers, and Michele Rossi, for consistent support during the editing process. A very special thanks is extended to the College of Business and Marketing Program at James Madison University for providing assistance. The *Advances in Electronic Marketing* book project would not have been possible without the hard work and dedication of the contributing authors and reviewers. All chapters appearing in this book were reviewed by the external reviewers through a double-blind review process. The external reviewers did not know the authors' names or affiliations. The chapter selection process was quite competitive and went through several revisions based on comments provided by reviewers and editors. Although we evaluated 33 chapters (70 authors) for the book, only 16 chapters (34 authors) were accepted for final inclusion.

Irvine Clarke III and Theresa B. Flaherty
James Madison University, USA

Section I:

Buyer Behavior
of Online Consumers

Chapter I

Attracting and Retaining Online Buyers:
Comparing B2B and B2C Customers

Eileen Bridges, Kent State University, USA

Ronald E. Goldsmith, Florida State University, USA

Charles F. Hofacker, Florida State University, USA

Abstract

This chapter addresses similarities and differences in e-commerce needs for customers in business-to-business (B2B) and business-to-consumer (B2C) marketplaces. We discuss how and why customers are attracted to online buying in general and to a supplier in particular for each of these types of markets. We further compare the characteristics of customers who choose to buy online with those who prefer to continue with more traditional means of purchasing, providing some possible reasons for observed differences. The customer's online experience may influence both satisfaction and buying behavior, so we address the antecedents of the experience, including Web site design and the nature of customer involvement with the site. We note the importance of Web site efficacy (usefulness and ease of use) as well as experiential elements of online shopping in customer satisfaction and retention and make specific recommendations for marketing managers in firms offering a Web presence.

Introduction

The emergence of e-commerce as a way of doing business has created an environment in which the needs and expectations of business customers and consumers are rapidly changing and evolving. This situation presents marketing managers with the challenge of ascertaining which elements of marketspace are new and how much continuity can be retained from the past. Some marketers apparently believe that it is enough to offer a Web site, maintaining a superficial appearance that the firm is progressive, or they ignore the Web altogether, possibly making use of digital technology to support existing business plans. Others take the opposite tack, saying that everything is changing and that nothing can remain the same (e.g., Feather, 2000; Murphy, 2000). A more balanced view proposes that people are basically the same but that new technologies are changing many of the ways customers shop and buy – thus, many businesses must overhaul their operating models to create digital strategies that meet changing needs and preserve competitiveness (Downes & Mui, 1998; Wind, Mahajan & Gunther, 2002). Our position is consistent with this balanced view. We believe that while much is changing, many fundamentals remain. Thus, we suggest ways that managers can use well-grounded concepts from consumer behavior and marketing theory, adapting them to new technologies.

We begin by considering characteristics of the most innovative online buyers and how marketers might best attract these individuals to Web sites. Quite a lot has been learned in the past few years about online customer behavior. This knowledge should benefit e-marketers in their efforts to develop successful online marketing strategies. These topics are important because now that many investors have been burned by dot-com business failures, they are unwilling to provide funding to start-ups that do not have clear business strategies in place and excellent indicators of potential success. In addition, competition is fierce, so competitive advantage is even more important on the Web than it is for an industrial sales representative calling on a customer or for a retail store in a neighborhood shopping mall. We propose that the more managers know about customers and their expectations of, as well as reactions to, e-commerce activities, the better these online vendors will be able to attract and to retain consumers. Thus, the objectives of this chapter are to discuss aspects of buyer behavior that offer strategic insights to e-commerce managers who wish to attract new buyers, satisfy, and retain them.

Who Buys Online?

"The ultimate objective of any given marketing strategy should be to attract, satisfy, and retain customers" (Best, 2004, p. 15). Thus, a discussion of online customer behavior from the management perspective should address this objective. Attracting new customers holds dual relevance. First, for any innovative good or service, the first buyers provide revenue needed to pay for research, development, and launch costs involved in bringing a new product to the market. Marketers must generate positive revenue streams quickly to grow their businesses. Moreover, it is not possible to retain all buyers as customers, so it is necessary to acquire new customers continuously to replace those who leave. Second, new customers are important because, in addition to an initial purchase, they bring the potential for lifetime value, which is the stream of profits that accrues during a customer's relationship with the firm. By itself, attracting new buyers is not enough; firms need to retain at least a portion of first-time purchasers to remain viable. Loyal satisfied customers have value for the firm because they may increase their spending over time, and they tend to spread positive word-of-mouth (via e-mail, etc.) about the company, attracting other buyers (Reichheld & Schefter, 2000). Finally, the earliest buyers may also provide valuable feedback regarding their experience with the Web site, facilitating site improvements.

How are companies using an online presence to reach potential and current customers? Liu et al. define online business as "the buying and selling of goods and services where part, if not all, of the commercial transaction occurs over

Figure 1. Overview of online buyer attraction, satisfaction, and retention

	TYPE OF INTERACTION	
	Business-to-Business (B2B)	Business-to-Consumer (B2C)

Attraction	• Marketing mix
	• Customer demographics
	• Convenience, efficiency
	• Use of digital data transfer
	• Availability of specific info
	• Documented quality
	• Support services
	• Warranty
		• Marketing mix
		• Consumer demographics
		• Consumer psychographics
		  (e.g., innovativeness)
		• Positive attitude toward
		  online buying
		• Prefer online functionality
		• Prefer online experience

Satisfaction	• Lower prices
	• On time, correct delivery
	• Excellent communication
	• Met expectations
	• Lack of problems or
	  opportunistic behavior
		• Positive online experience
		• Outcome meets expectations
		• Reduction of functional,
		  financial, social, and
		  psychological risk
		• Recovery in case of failure

Retention	• Repeated satisfying
	  experience
	• Consistent delivery
	• Convenient rebuy
	• Provide added value
	• Customer relationship
	  management (CRM)
		• Repeated satisfying
		  experience
		• Consistent delivery
		• Post-sale support
		• Frequent Web site updates
		• Loyalty programs and CRM
		• Building of trust

an electronic medium" (1997, p. 336). At the time of their article examining the Fortune 500, although 322 of the firms maintained home pages on the Web, only 131 of these provided for online business. Of course, these numbers are much higher now, but the motivations for making use of an electronic marketplace still hold. They suggest that by providing improvements in communication and information processing the Web greatly reduces the costs of market coordination and improves efficiency. Because potential buyers can compare offerings more easily, electronic markets should promote price competition as well as product differentiation. However, in the case of B2C e-commerce, research suggests that price dispersion is persistent due to differences in service offerings among e-tailers, market characteristics such as number of competitors, and possibly such factors as brand name and online trust (Pan, Ratchford & Shankar, 2002).

The results of Liu et al. (1997) indicate that presence of a home page and revenues are significantly related – this makes sense because the large majority of the home pages they observed were designed to introduce new products and provide overview information about the companies. While smaller sized firms are apparently more interested in direct selling and generation of revenue, larger firms' Web sites focus on communication activities. This suggests that while larger firms are more likely to try to build awareness upon entering an electronic marketplace, smaller firms may try to move directly into selling. The latter strategy may neglect the need to communicate and assist potential customers in developing familiarity with a firm and its products – this may be part of the reason so many smaller firms failed in the dot-com bust of 2000.

Our essential topics for the chapter follow from the need for an e-marketer to attract new customers to a Web site, satisfy, and retain them. Although many of the critical issues are the same in the B2B and B2C marketplaces, customer needs and behaviors differ for each of these types of markets. Thus, we address each separately in this chapter, drawing on both theoretical and applied research that has become available in the past few years. Our observations are summarized in Figure 1 for convenient reference.

Characteristics of B2B Online Buyers

In their analysis of the buying process for tangible, customized products, Gattiker, Perlusz, and Bohmann (2000) suggest that potential B2B online customers gather information needed for decision making and process it, taking into account situational factors and available product attributes. Further, individual buyers' demographic characteristics, cultural backgrounds, attitudes toward technology, and economic factors influence the decision-making process. For instance, they note that women are less likely than men to spend time on the Web and more likely to value interpersonal communication. Younger respondents are more likely than are older respondents to spend time on the Web and to have made at least one purchase online. More educated respondents are more likely to spend time on the Web than are less educated respondents. Outcomes, specifically Web beliefs and behaviors, depend on Web usage patterns, ability to test and purchase products online, type of information available, and usage situation.

While all customers might be expected to evaluate certain product attributes, particularly as their needs depend on the planned usage of the product, B2B purchasing agents differ from individual (B2C) shoppers (Gattiker et al., 2000). In particular, corporate buyers are more concerned than are B2C consumers with obtaining specific information, such as delivery conditions and pricing options. In addition, corporate agents may require documented quality and postsale support because of company policy. Other features that may be important include presale support, availability of postsale on-site service, and terms of replacement (e.g., lead times). Gattiker et al. (2000) further suggest that some differences between online and off-line shopping may occur for both B2B and B2C customers. For instance, extrinsic product attributes, including price and brand name, recommendations from others, and warranty, may be more important online than in a brick-and-mortar retail store where other product features may be more readily evaluated.

As Web sites began to provide more of the information necessary for B2B transactions to occur, this marketplace grew at a phenomenal rate, far surpassing B2C online revenues. According to E-Stats (2002, 2003), an online publication of the U.S. Department of Commerce, B2B e-commerce in the U.S. totaled $913 billion in 1999, $997 billion in 2000, and estimates suggested it would be $995 billion in 2001. The lower figure for 2001 may be due in part to a leveling-out of B2B online trade but may also be reflective of the general economic downturn that began in 2001. The publication also notes that "e-commerce outperformed total economic activity in three of four major economic sectors measured between 2000 and 2001" (2003, p. 1). Thus, e-commerce is very healthy in sectors that sell primarily to other businesses.

Some industries are more inclined to utilize e-commerce than others. E-Stats (2003) notes that 68% of all manufacturing e-shipments occur in only five industries, including transportation equipment, beverage and tobacco, electrical equipment, appliances, and components. Merchant wholesaler e-sales were concentrated in only three industry groups with drugs and druggist sundries, motor vehicles parts and supplies, and professional and commercial equipment and supplies explaining 64% of the total. There may be greater opportunities for increasing B2B e-commerce in industries that are not yet heavily represented.

Characteristics of B2C Online Buyers

In the early days of the Internet, the demographics of online consumers were skewed toward young well-educated male technophiles (Modahl, 2000). Today, the population of Internet users looks more like a representative national sample. In Table 1, adapted from Lenhart et al. (2003), we see a snapshot of current Internet users.

Table 1. Portions of groups that shop online

Gender	
Male	60%
Female	56%
Ethnic Group	
White	60%
Black	45%
Hispanic	54%
Education	
High School	45%
Some College	72%
College Graduate	82%
Annual Income	
<$30,000	38%
$30,000-$50,000	65%
$50,000-$75,000	74%
>$75,000	86%
Age	
18-29	74%
30-49	67%
50-64	52%
65 and over	18%

These data are consistent with information regarding B2B online shoppers in that men, more educated, and younger people are more likely to shop online. Such demographic differences are also declining in both cases.

E-Stats (2002) notes that e-commerce represents a relatively small share of the B2C marketplace as compared to economic activity in the B2B sectors. Total B2C online commerce was $40 billion in 1999, $65 billion in 2000, and $71 billion in 2001 (E-Stats, 2002, 2003), including services and retail trade. Thus, although it is growing as a share of total e-commerce activity in the U.S., B2C represents only 4.2% of e-commerce in 1999, 6.1% in 2000, and 6.7% in 2001. Four industry groups account for half of e-revenues in B2C services, including travel arrangement and reservation, publishing (including software), securities and commodities contracts intermediation and brokerage, and computer systems design and services (E-Stats, 2003). The information industry is an area of services having particularly strong growth, but it is not yet well represented online. Thus, this may indicate an opportunity for e-commerce activity.

Retail e-sales are concentrated in only two groups that account for over 90% of this marketplace; these are nonstore retailers (including brick and click, catalog, and pure play businesses) and motor vehicle and parts dealers. Retail e-sales grew 22% between 2000 and 2001 as compared to total retail sales growth of only 3% (E-Stats, 2003). Merchandise categories having the highest percentage of online sales include books and magazines and electronics and appliances. This is consistent with previous research, which suggests that books and CDs are common items for initial online purchases (Florsheim & Bridges, 1999). Other product categories are more likely to be purchased by more experienced online shoppers and thus may have greater potential for e-tailer entry and growth.

How do consumers begin to buy online? This question can be placed in a larger context of online consumer behavior, defined as "any Internet-related activity associated with the consumption of goods, services, and information." Further, "Internet consumption includes (1) gathering information passively via exposure to advertising, (2) shopping, which includes both browsing and deliberate information search, and (3) the selection and buying of specific goods, services, and information" (Goldsmith & Bridges, 2000, p. 245). We can expand this definition by suggesting that online consumer behavior also includes reacting to postsale activities offered by Web merchants. Consumer loyalty may be associated with these postsale interactions. We note that for any new good, service, or idea, some B2C buyers purchase earlier than do others following the launch of an innovation. Thus, it may be appropriate to study the purchase behavior of online buyers using a model for diffusion of innovation, particularly to improve understanding of how they might be attracted to an e-commerce vendor.

Attracting Buyers Online

Potential customers who become aware of a new product have several possible reactions. First, they may delay action. Diffusion theory (Rogers, 1995) categorizes consumers into five adopter types based on how long they delay trying a new product: innovators, early adopters, early majority, late majority, laggards. Innovators have the quickest time-of-adoption and laggards the slowest. *Ceteris paribus*, this theory presumes that at some point all potential customers try the new product. Based on this trial, they may become active rejecters of the product. This may be because the product failed to satisfy their needs and wants, so they search for something else. If the new product is satisfactory or superior to existing products, they may decide to adopt it and become owners/users. Subsequently, some of these adopters may remain loyal to the new product; in this instance, they will become loyal and repeat buyers. If this process occurs in the context of e-commerce, the customers become loyal purchasers from the Web site. It is also possible that they may be dissatisfied with the Web site and avoid buying there again; thus, dissatisfaction with the Web site may carry over into dissatisfaction with the product itself.

Thus, the first task of an online merchant is to attract potential buyers to its Web site. Why would a customer choose to order online if alternative channels are available? For both B2B and B2C buyers, these alternatives can include retail stores or sales representatives, the telephone, other electronic media, or using a catalog to place mail orders. Online marketspace presents many advantages (Goldsmith, 1999; Hofacker, 2001; Xue, Harker & Heim, 2000), including:

- 24/7 product availability
- wider product selection, availability of niche items
- low prices
- prompt delivery
- access to customer service
- personalized treatment
- opportunities for two-way communication
- convenience of shopping at home or office
- privacy
- no pressure from salespeople
- ease of search and comparison
- low switching costs
- benefits of collaborative filtering

Nevertheless, there are also potential disadvantages:

- loss of privacy or personal information
- delayed gratification owing to slow delivery
- errors in order fulfillment

- potential for fraud
- negative interactions with online merchant
- inability to inspect or experience the product prior to purchase
- low fidelity of the online shopping experience
- cost of computer ownership and network connection
- slow modems and site loading times

Potential customers must evaluate the benefits and risks associated with online purchasing before they opt to try it. The nature of these issues differs somewhat for B2B and B2C marketspaces. As noted earlier, business customers must consider practical factors related to meeting requirements efficiently. On the other hand, researchers in online consumer behavior have studied a variety of factors that predispose consumers to try online buying, including those related to usefulness and ease of use but also those related to enjoyment of the experience. Individual attitudes and personal predispositions may influence reactions to the possibility of ordering on the Web and to the Web experience itself. For instance, Florsheim and Bridges (1999) observe that customers who are insecure and feel the need to ask questions are less likely to shop online. Such customers report that they prefer the human contact available when placing an order by phone to the anonymity of the Internet, despite its potential for improved consistency and convenience. Further, their finding that men are more willing than women to give up human contact is consistent with other studies that suggest men are also more likely to order online (Goldsmith & Bridges, 2000).

Innovative B2B Firms: Moving Purchasing to the Web

Before discussing the motivation for B2B customers to move purchases to the Web, it will be helpful to think about some of the ways that business purchasing differs from consumer purchasing. Compared to an individual consumer, business buyers tend to purchase in larger volumes, and sellers tend to have fewer but larger customers. Promotional activities are more likely to be personal and interactive rather than impersonal and one-way as compared to mass-media advertising and large-scale promotions appropriate for consumer marketplaces. Purchasing options might include the possibility of protracted negotiation, reciprocity, or leasing; thus, they are more complex than typical consumer purchases. Unlike individual consumers, businesses are quite used to direct channels as B2B purchases have traditionally utilized such approaches. Thus, the Internet represents another mechanism for facilitating the supply chain, along with 800 numbers, catalogs, and direct sales forces.

Business purchasers are typically motivated to find least-cost solutions to purchasing problems, subject to meeting minimum objectives. When they place orders repeatedly for the same products and these are filled through a standard process, there is no need to interact with an order taker or sales representative to make a purchase decision. Although such personal interaction might be enjoyable for some buyers, firms face pressure to improve efficiency and are constantly looking for ways to reduce waste. Of course, a number of personal selling encounters might be required before two firms agree to engage in automatic e-commerce But then, a routine rebuy would certainly be flagged as the type of purchase that might be completed automatically, reducing both time requirements and the potential for human error. Thus, B2B buyers tend to move online more easily for routine than nonroutine purchases. With the help of inventory management software tied to an extranet, some of these routine orders can even be placed in a fully automated way, eliminating the need for employee time to be used.

When a new type of purchase is considered, the story is different. Following the appearance of an innovation in the marketplace, potential B2B customers begin to consider their firm's need for it. They initiate the process of search and evaluation, possibly aided by information available on one or more Web sites. However, particularly if the product requires a large investment (such as capital equipment), it may be necessary to involve purchasing agents, sales representatives, engineers, and other personnel in a rather complex purchase process. The firm's online interaction may still be a part of the process because it can provide information, reduce search costs, and facilitate relevant communication and postsales follow-up activities. Companies use the Web for many types of postpurchase support, ranging from reviewing technical specifications and ordering spare parts to participating in user group discussion forums. When the purchase is routine, but the timing is not systematic, B2B e-hubs or online marketspaces implementing one of the auction types described below can perform a valuable matchmaking function (Kaplan & Sawhney, 2000).

Wilson and Abel (2002) discuss e-hubs as they reflect on issues that a firm must address before it can successfully launch a B2B online business. These ideas are useful in better understanding online buyers. They suggest that B2B customers hope to gain such benefits as improved communication with the supplier, enhanced productivity with just-in-time delivery, collaboration in product and process design, improved information at lower cost, and administrative/operational cost reductions. They also note that B2B online marketplaces may be designed in several ways. *Forward auctions*, which tend to sell excess inventory at low prices involve buyers bidding for this inventory. *Reverse auctions* work with sellers bidding to arrange long-term contracts on standardized products. In addition to the direction of the auction, e-hubs can either involve single firms and single buyers or aggregate multiple buyers, multiple sellers, or both. Buyers must evaluate these possible options in deciding which type of B2B purchasing arrangement best meets firm needs, but the returns to scale make participation in these electronic markets worthwhile for many companies (Kaplan & Sawhney, 2000).

It is instructive to consider how the adoption of just-in-time inventory management is tied to use of the Internet as a direct sales channel. Firms are motivated to keep inventory small for a number of reasons: (1) reducing the cost of the capital tied up in inventory; (2) reducing costs associated with breakage, insurance, storage; and (3) enhancing the ability to quickly adapt to changing market conditions. However, the use of frequent small deliveries requires careful coordination between buying and selling firms. The Internet and other IT facilities (such as enterprise resource planning) allow companies to automate coordination and administration, even between two independently managed companies. Careful coordination is therefore increasingly important within the supply chain—firms engaging in long term relational exchange must coordinate packaging decisions, delivery vehicle and warehouse configurations, and storage requirements right from the product design phase. But in order to substitute information for inventory, purchasing must adopt e-commerce exchange, permitting the cost benefits and improved time flexibility.

Large firms are more likely than small firms to design and implement successful B2B online commerce. This makes sense because larger firms have more to gain from such an investment, which requires installation of automated supply chain management and inventory control (such as an enterprise management system). Min and Galle (2003), in a study designed to observe differences between adopters and nonadopters of e-purchasing, find that larger firms having a greater number of purchasing employees are more likely to adopt. Further, firms in more information intensive industry sectors are more likely to purchase online than those in less information intensive sectors. Thus, firms that can best benefit from e-purchasing are those that stand to gain the most from digital data transfer, enhanced supply chain efficiency, real-time information regarding product availability, inventory level, shipment status, and production requirements.

In short, although individual characteristics, such as personality, may be expected to drive B2C Internet adoption, structural and economic factors drive businesses to switch to B2B e-commerce. Such factors include specific inventory requirements, newness of the type of purchase, requirements for engaging in spot markets, benefits that might accrue from having a single hub that aggregates suppliers, sales volume, firm size, and competition in the marketplace.

Innovative B2C Consumers: The Role of Attitudes and Personality

Among the many constructs used to describe, explain, and influence online consumer behavior, attitudes take a central position. This is because attitudes have traditionally been widely used to study consumer behavior, they provide diagnostic information for marketing management, they are easy to describe and explain, and they can be influenced by marketing strategy (compare Modahl, 2000). Goldsmith and Bridges (2000) observe that positive attitudes toward buying online are associated with online purchase of textbooks. Specifically, students who have purchased textbooks online are more likely to think that the Internet is easy, quick, and safe to use than are students who have not. Further, they are more confident that orders will be filled accurately and promptly; they feel that online merchants have good return policies; and they believe that buying online offers better value than does a brick-and-mortar alternative. Goldsmith, Bridges, and Freiden (2001) confirm these findings with adult consumers and further observe that Internet innovators are more likely than other consumers to buy online.

Goldsmith (2000) measures the attitudes of innovative online buyers using the Domain Specific Innovativeness Scale or DSI (Goldsmith & Hofacker, 1991) formatted for the topic of online buying (compare Goldsmith, 2001). This six-item self-report scale uses a Likert response format to measure a consumer's innovativeness concerning a particular product category (Goldsmith, d'Hauteville & Flynn, 1998). The results suggest that innovative online buyers are more likely to have made online purchases; further, online buying innovativeness is positively related to amount of online buying for those who had purchased this way. Moreover, comparing the attitudes of early adopters, early and late majorities, and laggards, the more innovative consumers are more likely to view online shopping as quicker, cheaper, and more fun than traditional shopping. They also think it is safer to buy online and are more confident of their ability to buy online. Goldsmith and Lafferty (2002) replicate these findings in a later study, where they also observe that innovative online buyers purchase more online and believe themselves to be more knowledgeable about online buying. Moreover, higher levels of Internet innovativeness are associated with a greater reported likelihood of future online purchases.

Flynn and Goldsmith (2001) provide evidence suggesting that Internet knowledge leads to positive attitudes that are subsequently linked to online buying. This finding is supported by a study of online apparel buying (Goldsmith & Goldsmith, 2002). This analysis shows that online apparel buyers purchase nonapparel products online more often than do those who do not purchase apparel online. That is, those who buy apparel online also buy other products online as well, suggesting a general pattern of buying online. Consumers who buy apparel online also embrace the attitudes that buying online is safe, quick, and fun, and they are more confident in their ability to buy online than are nonbuyers. They do not buy apparel in general more than do the nononline buyers, suggesting that they tend to shift their buying from stores and catalogs to cyberspace. They are more innovative and more knowledgeable than are the nonbuyers about the Internet.

Holbrook and Hirschman (1982) point out how consumers receive two different types of benefits from shopping. There is a utilitarian benefit due to finding the product that one is looking for and receiving the functional benefits thereof. In addition, at least for some consumers, there is the benefit of the shopping experience itself. The pleasure of looking in shop windows, being greeted by and socializing with the store help, and experiencing the sensory aspects of store atmospherics can contribute to the satisfaction derived from the shopping experience itself. Analogously, studies of innovative online buying show that consumers may be motivated to shop by both utilitarian and hedonic dimensions of the retail atmosphere (Childers et al., 2001). These authors note that until recently shopping on the Internet has been viewed as providing primarily utilitarian value. However, its "immersive, hedonic aspects" are now more appreciated, creating what they term the "webmosphere" (p. 511). Childers et al. (2001) use the Technology Acceptance Model (TAM) described by Davis, Bagozzi, and Warshaw (1989) to understand how technology-based characteristics of the Web influence its adoption. Specifically,

Childers et al. (2001) consider the Web site's utilitarian aspects, which may be described as usefulness (or outcome) and ease of use (or process), as well as its hedonic aspects. The latter provide enjoyment (also a process dimension of the shopper's online experience), which is separate from any performance results. Although Davis et al. (1989) find that usefulness is the primary determinant of a decision to adopt technology, Childers et al. (2001) anticipate that process dimensions of the online experience and enjoyment in particular may play a greater role than previously thought in a consumer's decision to shop online. Specifically, they postulate that context may differentiate the importance of different antecedents of technology adoption – their results indicate that usefulness is more related to shopping online when the goals are instrumental, while enjoyment is more related when the shopping is hedonic.

Chen, Gillenson, and Sherrell (2002) suggest that there are interesting similarities between the TAM, mentioned above, and innovation diffusion theory discussed earlier in the chapter (see Attracting Buyers Online). For instance, one driver of diffusion is relative advantage of the innovation, which is comparable to TAM's usefulness construct. Similarly, complexity in diffusion theory is inversely related to ease of use in TAM. The authors also investigate compatibility of the innovation with existing systems and values – this concept comes from diffusion theory but is not included in the TAM formalization. They did not include two other drivers of innovation, trialability and observability, which are often included in diffusion research but which they did not expect to relate to adoption of online shopping. Their findings suggest that compatibility, usefulness, and ease of use are important drivers of a consumer's decision to adopt online shopping and that they will increase sales and loyalty. They note that compatibility and usefulness are highly correlated, which is not surprising because both indicate the outcome value to the consumer. The authors recommend that Web marketers should increase consumers' enjoyment of their sites because this is found to motivate increased use. However, their results do not indicate that this will lead to increases in purchasing.

What Web site design elements will lead to increased buying? Childers et al. (2001) demonstrate the dual importance of usefulness and enjoyment to the Web customer, but how can the site design be improved to achieve these objectives? They suggest that flexibility in navigation, convenience, and the substitutability of the Web site visit for personal examination of the product are critical characteristics. Thus, the online experience must be intrinsically enjoyable while it offers some improvement over the physical retail store. Childers et al. (2001) indicate the need to understand how elements of physical retail store design (external architecture, interior design and layout) can be translated into more effective Web site design. They comment that "very little is known about how the design characteristics of interactive shopping sites affect online purchase behavior and other usage indicators" (p. 529). These design elements might include structural attributes (e.g., frames, graphics, text, pop-up windows, etc.), media (graphics, text, audio, etc.) and site layout (organization of product offerings).

Another aspect of the online experience that could potentially influence buying behavior is the concept of flow. Hoffman and Novak (1996) define flow as "the state occurring during network navigation which is: (1) characterized by a seamless sequence of responses facilitated by machine interactivity, (2) intrinsically enjoyable, (3) accompanied by a loss of self-consciousness, and (4) self-reinforcing" (p. 57). These same authors propose that "creating a commercially compelling website depends on facilitating a state of flow" (p. 66). This is corroborated by Goldsmith et al. (2001), who find that the reported amount of online buying is greater for Internet users that experience two key elements of flow – confidence and fun. Korzaan (2003) also tests for the impact of flow on purchase, finding that being in flow positively influences a customer's intention to buy online.

In summary, marketers will be most successful in attracting consumers to their Web sites if they first are diligent in discovering the specific needs that their respective customers seek to satisfy. Next, they must craft products so that they will satisfy these needs. The Internet is one channel among many through which merchants can contact and interact with customers. A Web site should be integrated into existing and future channels so that they are mutually supportive. The Web site itself should be easy to use and useful to consumers who shop and want to buy. The earliest buyers, those most likely to visit the Web site, view e-commerce as a convenient, quick, and safe method of shopping. The Web site should reflect these attitudes. Moreover, some consumers buy online because they are interested in the product category and are looking for products unavailable to them via other channels; they lack the time, energy, or ability to physically shop, they seek to save money, or they enjoy the activity of online browsing and buying. Enhancing the aesthetic, hedonic, and entertaining aspects of Web sites will make them attractive to many innovative consumers. However, attracting potential buyers to a Web site and encouraging them to purchase is only the beginning of an online relationship.

Retaining Online Buyers

Attracting new buyers to online sites is only part of the job of the e-marketer. The three basic uses consumers make of the cyber-marketplace are to get information, to purchase goods and services, and to obtain postsale support; thus, communication, online sales, and customer service are necessary (Griffith & Krampf, 1998). Retaining buyers is a function of competitive activity and customer satisfaction. Intense competition and low switching costs characterize the computer-mediated environment, so consumers have ample opportunity and motivation to switch vendors (Xue et al., 2000). Consequently, customer satisfaction with a Web site is critical to maintaining loyalty. Because this issue is so important, quite a bit of research has been done on satisfaction with e-commerce Web sites.

Customer relationship management (CRM) systems have often been recommended as a means of attracting and retaining customers. Because it is very convenient to collect customer data (including demographics and buying habits) online, and it is also easy to interact and continue a relationship online, it is often assumed that adopting an e-CRM system is a good way to draw together all of the elements of marketing strategy to obtain customer satisfaction, sales, and profit. However, recent research shows that although product customization ability is important, and the presence of some e-CRM features is positively related to customer satisfaction with a Web site, e-CRM activities are unrelated to success in attracting and retaining buyers (Feinberg & Kadam, 2002).

Retaining B2B Buyers

One key to retaining B2B buyers in an online environment lies in using the Internet to facilitate collaboration between supply chain partners (Downes & Mui, 1998). Because it is difficult to build and maintain trust in the online environment, a critical success factor in online B2B relationships is the absence of opportunistic behavior by either of the firms (Williamson, 1994). Our discussion of B2B purchasing assumes the use of relationship marketing ideals, which go beyond focusing on specific transactions and seek to optimize the lifetime value of each customer.

Many of the characteristics of B2B relationships in traditional marketplaces continue to be of interest when building online relationships. For instance,

Freytag and Bridges (2003) observe that more dynamic marketplaces, characterized by changing relationship structures, require the firms involved to invest both financially and otherwise in order to obtain continuing relationships. Because online markets are so new and involve innovative players, they are inherently dynamic; thus, they would be expected to require continuing investment in order to be successful. Freytag and Bridges also find that if a marketplace is unbalanced, that is, that the relative size and strength of buyers and suppliers are unequal, this influences the best choice of marketing strategy. For instance, if suppliers have greater marketplace intensity than do buyers, sellers must invest more in order to build B2B relationships. Due to the high rate of change, there is an inherent lack of balance in online markets, consistent with this idea that suppliers must invest in the relationships.

Finally, as more firms compete in a particular marketplace, the competition is increasingly value-based (compare Porter, 1985). Thus, as the popularity of online B2B marketplaces grows, sellers may be expected to compete more strongly through differentiation that provides added value. Further, by integrating customers into its own systems and procedures, a firm may create switching costs that act as a barrier to leaving the relationship (compare Downes & Mui, 1998). The idea of customizing the product to provide added value and to lock in customers is consistent with reaping the benefits of customer lifetime value. Thus, an online relationship strategy can create bonds between companies to the benefit of all involved.

Retaining B2C Buyers

One key to retaining B2C consumers is to change the Web site continuously, so it appears fresh and alive and to utilize timely specials that appear at unpredictable intervals. In addition, the firm should strive to add value for consumers by offering product usage tips, specifications, manuals, parts lists, glossaries, histories, and so forth. For instance, Stolichnaya Vodka offers cocktail recipes on its site, while Molson Brewers at one time had contest winners wear a Web camera as they pub-hopped. In general, the opportunity to interact with other users of the product through games, chat rooms, and idea exchanges can lead to enhanced flow within the site and lead to more a more favorable attitude towards the site and the sponsoring company.

In some markets, B2C online sales grew very quickly but were unable to sustain the aggressive pace required. This may occur because service quality erodes during a period of rapid growth, possibly due to a lack of infrastructure or experience. Thus, the firm is vulnerable to competition and may find that it needs to drop prices just when it can least afford to (Oliva, Sterman & Giese, 2003). The quality of service the consumer receives is crucial to customer retention in an online B2C setting. Thus, it is important to keep prices at a level where the Web site can attract a sufficient number of customers to make a profit while continuing to provide quality service.

Research into the failure of dot-com businesses suggests that, although timing and industry evolution are important in sustained buying, other factors may come into play as well. Specifically, Thornton and Marche (2003) observe that B2C online businesses often did not have adequate contingency planning to allow for unexpected events. Thus, consumers experienced poor quality when failures occurred in the buying process. Further, many businesses were started by individuals lacking in specific industry knowledge, who were therefore unable to foresee customer needs or make necessary changes in a timely manner. Trust may also be a critical success factor. Gefen, Karahanna, and Straub (2003) provide evidence that consumers are more likely to trade with online businesses that they trust than with those that do not earn their trust. The use of e-CRM depends on the ability to collect personal information online. All the e-CRM in the world will not win a buyer if it is perceived that the company is benefiting from the inappropriate provision of personal information, inconveniencing or harming consumers.

Conclusions

Design of the Web site can be crucial in attracting, satisfying, and retaining customers. Hopkins, Raymond, and Grove (2003) summarize the relevant issues, which include site design and navigability, information and content, reliability, and interactivity. Because the Web site is an electronic servicescape, they coin the term *e-servicescape* and suggest that physical design aspects of service facilities be applied to Web sites. Based on Bitner's (1992) suggestions for servicescape design, they consider ambient conditions (atmospherics, context, design, entertainment value), spatial layout and functionality (site layout, navigability, and reliability), and signs, symbols, and artifacts (product information, text, and graphics). Their findings indicate that signs, symbols, and artifacts (information dimension) and ambient conditions (entertainment dimension) strongly influence customer attitude and purchase intention. Spatial layout and functionality features influence attitude but not intention to buy.

Korzaan (2003) recommends that Web site designers attempt to induce the state of flow in site visitors, in particular, those elements of flow associated with a sense of control, challenge, and stimulation. Specific ideas include eliminating error messages and dead links to allow for seamless navigation and increasing the interactive speed as far as possible. Further, information can be presented directly through comparison advertising, provided continuously in a banner or sidebar, or indirectly through graphical pop-up or floating advertisements. For the more experiential side of Web site design, Huang (2003) considers complexity (amount of information the site offers), novelty, and interactivity. Study results indicate that complex Web sites are perceived to be useful (but distracting), while interactivity is the key to creating flow, working through control, curiosity, and interest. Novelty facilitates flow by exciting curiosity but undermines hedonic Web site performance. We can add to these general recommendations three specific managerial suggestions for retaining online buyers.

First, the multidisciplinary field of human-computer interaction is the study of how people interact with computing technology (Nielsen, 2000; Olson & Olson, 2003). Experts in this field provide advice on the design and operation of Web sites to make them useful, usable, and aesthetically pleasing. While this recommendation may seem trivial, many businesses drive away customers after making an effort to attract them to Web sites by presenting them with clumsy, irritating, unhelpful, slow, and generally unfriendly places to shop. We now know that Web site design characteristics can greatly impact Web site usage (Montoya-Weiss, Voss & Grewal, 2003), and in fact, even subtle details like the choice of background graphic can influence purchasing (Mandel & Johnson, 2002). Of course, minimizing download time or providing enhancements during the waiting period are good ways to improve customer-Web site interactions (Weinberg, Berger & Hanna, 2003). A Web site that does not appear to be professional, that is, it features poor graphics, ill-chosen fonts, and sloppy and badly arranged sections, may lead potential customers to distrust it and turn to another merchant, especially if the poorly designed site is for a less well-known product. Competing with established online marketers means matching or exceeding their sites as a first step. The more involving the site through interactivity, the more positively it is perceived (Ghose & Dou, 1998). Moeller, Egol, and Martin (2003) might term an easy to use, useful, and entertaining Web site an *order qualifier*, that is, a characteristic that would "enable companies to stay in the battle with competitors" (p. 5). Other marketers might call an effective Web site a *must have*.

Second, e-commerce managers must cultivate a culture of service if they are to retain customers. They must recognize that a purchase is a process that does not end with the sale. Customers often want to be able to interact with their vendors before, during, and after the sale. They want information and advice. Merchants must strive to provide these services and emphasize promptness, helpfulness, and knowledge in dealing with customers. Whether customers buy online is positively influenced by ease and convenience and highly influenced by whether this behavior fits into their lifestyle (Becker-Olsen, 2000). Moreover, marketers should focus on building trust between themselves and customers (Wilhelm, 2003). Research into online buying shows that customers' trust in their vendors plays an important role in their continued relationships (Gefen et al., 2003). The basic principles of brand building should be applied in the online context. A strong brand based on customer satisfaction is often the means to build customer trust. One way to enhance this aspect of the online strategy is to sponsor the development of online communities in which customers provide product feedback, information to each other, and recommendations that both relieve the insecurities associated with online buying and encourage purchase (compare Hagel, 1999).

Third, e-commerce marketers should use the unique features of the Internet to customize their interactions with customers (Goldsmith, 1999). Personalization or customization is the tailoring of a business strategy to match the needs, characteristics, and behaviors of individual customers. As the online customer base grows, it is apparent that one driver of customer retention is the extent and success with which a merchant customizes each buyer's interactions with the Web site. While this concept is well known in CRM (Peppers & Rogers, 1993), many e-marketers are less successful than they should be in personalizing interactions because they do not match the extent of their effort (the costs in money and complexity) with the level desired by their customers to maximize profitability (Moeller et al., 2003). These authors suggest that to create customer value and retain customers profitably managers should optimize the entire customer interaction process as follows:

- Add variety to offerings so that each customer receives the desired value package, and integrate these offerings into the production and delivery system in a cost-effective manner.

- Develop a deeper understanding of customer needs so that the customized offering becomes an order winner, that is, it meets a customer's most critical needs. This is especially true in a B2B setting.

- Tailor business streams to provide value at least cost. "Smart customizers match their segmentation strategies with delivery mechanisms designed specifically to serve each segment profitably" (p. 5).

In summary, attracting and retaining customers online is much like accomplishing these tasks in other business settings. The old recommendations are still true. Know your customers, create value, and provide satisfaction. Cyberspace presents many challenges to marketers because it is a new setting with new complexities and problems. Managers can cope with these, however, by using research results to learn how customers behave, what they need and want, and how they interact with the new media. These efforts should be directed to improving three crucial aspects of the online buying experience, including the interface, the encounter, and fulfillment quality (Chang, 2000). Online marketplaces also offer substantial advantages not available in other types of markets. These include the ability to provide product information conveniently, place orders 24/7, and unobtrusively gather customer database information for use in customer relationship management. Thus, over time the Internet will assume a place alongside other types of marketplaces, customers will shop and buy successfully, and businesses will thrive by adapting to this new reality.

References

Becker-Olsen, K.L. (2000). Point, click and shop: An exploratory investigation of consumer perceptions of online shopping. *Proceedings of the American Marketing Association Summer Educators' Conference, 11*, 62-63.

Best, R. (2004). *Market-based management* (3rd ed.). Upper Saddle River, NJ: Prentice Hall.

Bitner, M.J. (1992). Servicescapes: The impact of physical surroundings on customers and employees. *Journal of Marketing, 56*(2), 57-71.

Chang, T.Z. (2000). Online shoppers' perceptions of the quality of internet shopping experience. *Proceedings of the American Marketing Association Summer Educators' Conference, 11*, 254-255.

Chen, L.D., Gillenson, M.L., & Sherrell D.L. (2002). Enticing online consumers: An extended technology acceptance perspective. *Information and Management, 39*(8), 705-719.

Childers, T.L., Carr, C.L., Peck, J., & Carson, S. (2001). Hedonic and utilitarian motivations for online retail shopping behavior. *Journal of Retailing, 77*(4), 511-535.

Davis, F.D., Bagozzi, R.P. & Warshaw, P.R. (1989). User acceptance of computer technology: A comparison of two theoretical models. *Management Science, 5*(8), 982-1003.

Downes, L., & Mui, C. (1998). *Unleashing the killer app: Digital strategies for market dominance.* Boston: Harvard Business School Press.

E-Stats. (2002). Washington, DC: U.S. Census Bureau. Retrieved October 4, 2004, from *http://www.census.gov/estats*

E-Stats. (2003). Washington, DC: U.S. Census Bureau. Retrieved October 4, 2004, from *http://www.census.gov/estats*

Feather, F. (2000). *Future consumer.com.* Toronto: Warwick Publishing.

Feinberg, R., & Kadam, R. (2002). E-CRM Web service attributes as determinants of customer satisfaction with retail Web sites. *International Journal of Service Industry Management, 13*(5), 432-451.

Florsheim, R., & Bridges, E. (1999). Real people, real information: Is something missing in the internet marketing of consumer services? *Proceedings of the 1999 SERVSIG Services Research Conference: Jazzing into the New Millennium, 1*, 3-10. Chicago, IL: American Marketing Association.

Flynn, L.R., & Goldsmith, R.E. (2001). The impact of internet knowledge on online buying attitudes, behavior, and future intentions: A structural modeling approach. In T.A. Sutter (Ed.), *Marketing advances in pedagogy, process, and philosophy* (pp. 193-196). Stillwater, OK, Society for Marketing Advances.

Freytag, P.V., & Bridges, E. (2003). The importance of market conditions in B2B relationships. *Proceedings of the Australia and New Zealand Marketing Academy Conference*, Adelaide, Australia: University of South Australia, 2415-2421.

Gattiker, U.E., Perlusz, S., & Bohmann, K. (2000). Using the internet for B2B activities: A review and future directions for research. *Internet Research: Electronic Networking Applications and Policy, 10*(2), 126-140.

Gefen, D., Karahanna, E., & Straub, D.W. (2003). Trust and TAM in online shopping: An integrated model. *MIS Quarterly, 27*(1), 51-90.

Ghose, S., & Dou, W. (1998). Interactive functions and their impacts on the appeal of Internet presence sites. *Journal of Advertising Research, 38*(March-April), 29-43.

Goldsmith, R.E. (1999). The personalized marketplace: Beyond the 4Ps. *Marketing Intelligence and Planning, 17*(4), 178-185.

Goldsmith, R.E. (2000). How innovativeness distinguishes online buyers. *Quarterly Journal of Electronic Commerce, 1*(4), 323-333.

Goldsmith, R.E. (2001). Using the domain specific innovativeness scale to identify innovative internet consumers. *Internet Research: Electronic Networking Applications and Policy, 11*(2), 149-158.

Goldsmith, R.E., & Bridges, E. (2000). E-tailing versus retailing: Using attitudes to predict online buying behavior. *Quarterly Journal of Electronic Commerce, 1*(3), 245-253.

Goldsmith, R.E., & Goldsmith, E.B., (2002). Buying apparel over the Internet. *Journal of Product and Brand Management, 11*(2), 89-102.

Goldsmith, R.E., & Hofacker, C.F. (1991). Measuring consumer innovativeness. *Journal of the Academy of Marketing Science, 19*(3), 209-21.

Goldsmith, R.E., & Lafferty, B.A. (2002). Consumer response to Web sites and their influence on advertising effectiveness. *Internet Research: Electronic Networking Applications and Policy, 12*(4), 318-28.

Goldsmith, R.E., Bridges, E., & Freiden, J.B. (2001). Characterizing online buyers: Who goes with the flow? *Quarterly Journal of Electronic Commerce, 2*(3), 189-197.

Goldsmith, R.E., d'Hauteville, F., & Flynn, L.R. (1998). Theory and measurement of consumer innovativeness. *European Journal of Marketing, 32*(3/4), 340-53.

Griffith, D.A., & Krampf, R.F. (1998). An examination of the Web-based strategies of the top 100 U.S. retailers. *Journal of Marketing Theory and Practice, 6*(3), 12-23.

Hagel, J. (1999). Net gain: Expanding markets through virtual communities. *Journal of Interactive Marketing, 13*(1), 55-65.

Hofacker, C.F. (2001). *Internet marketing* (3rd ed.). New York: John Wiley & Sons.

Hoffman, D.L., & Novak, T.B. (1996). Marketing in hypermedia computer-mediated environments: Conceptual foundations. *Journal of Marketing, 60*(3), 50-68.

Holbrook, M.B., & Hirschman, E.C. (1982). The experiential aspects of consumption: Consumer fantasies, feelings, and fun. *Journal of Consumer Research, 9*(September), 132-40.

Hopkins, C.D., Raymond, M.A., & Grove, S.J. (2003, October). Designing the e-servicescape: Data-driven observations. *Proceedings of the American Marketing Association Frontiers in Services Conference.* College Park, MD, University of Maryland.

Huang, M.H. (2003). Designing Web site attributes to induce experiential encounters. *Computers in Human Behavior, 19*(4), 425-442.

Kaplan, S., & Sawhney, M. (2000). E-hubs: The new B2B marketplaces. *Harvard Business Review, 78*(3), 97-103.

Korzaan, M.L. (2003). Going with the flow: Predicting online purchase intentions. *Journal of Computer Information Systems, 43*(4), 25-31.

Lenhart, A., Horrigan, J., Rainie, L., Allen, K., Boyce, A., Madden, M., & O'Grady, E. (2003). The ever-shifting Internet population. *The Pew Internet and American Life Project.* Retrieved October 4, 2004, from *http://www.pewInternet.org/reports/pdfs/PIP_Shifting_Net_Pop_Report.pdf*

Liu, C., Arnett, K.P., Capella, L.M., & Beatty, R.C. (1997). Web sites of the Fortune 500 companies: Facing customers through home pages. *Information and Management, 31*(6), 335-345.

Mandel, N., & Johnson, E.J. (2002). When Web pages influence choice: Effects of visual primes on experts and novices. *Journal of Consumer Research, 29*(2), 235-245.

Min, H., & Galle, W.P. (2003). E-purchasing: Profiles of adopters and non-adopters. *Industrial Marketing Management, 32*(3), 227-233.

Modahl, M. (2000). *Now or never: How companies must change today to win the battle for internet consumers.* New York: Harper Business.

Moeller, L., Egol, M., & Martin, K. (2003). Smart customization. Retrieved October 4, 2004, from *http://www.portfoliomgt.org/read.asp?ItemID=1732*

Montoya-Weiss, M.M., Voss, G.B., & Grewal, D. (2003). Determinants of online channel use and overall satisfaction with a relational, multichannel service provider. *Journal of the Academy of Marketing Science, 31*(4), 448-58.

Murphy, T. (2000). *Web rules*. Chicago: Dearborn Financial Publishing.

Nielsen, J. (2000). *Designing Web usability*. Indianapolis: New Riders Publishing.

Oliva, R., Sterman, J.D., & Giese, M. (2003). Limits to growth in the new economy: Exploring the 'get big fast' strategy in e-commerce. *System Dynamics Review, 19*(2), 83-117.

Olson, G.M., & Olson, J.S. (2003). Human-computer interaction: Psychological aspects of the human use of computing. *Annual Review of Psychology, 54*, 491-516.

Pan, X., Ratchford, B.T., & Shankar, V. (2002). Can price dispersion in online markets be explained by differences in e-tailer service quality? *Journal of the Academy of Marketing Science, 30*(4), 433-445.

Peppers, D., & Rogers, M. (1993). *The one-to-one future: Building relationships one customer at a time*. New York: Currency Doubleday.

Porter, M.E. (1985). *Competitive advantage*. New York: Free Press.

Reichheld, F., & Shefter, P. (2000). E-loyalty: Your secret weapon on the Web. *Harvard Business Review, 78*(4), 105-114.

Rogers, E.M. (1995). *Diffusion of innovations* (4th ed.). New York: Free Press.

Thornton, J., & Marche, S. (2003). Sorting through the dot bomb rubble: How did the high-profile e-tailers fail? *International Journal of Information Management, 23*(2), 121-138.

Weinberg, B.D., Berger, P.D., & Hanna, R.C. (2003). A belief-updating process for minimizing waiting time in multiple waiting-time events: Application in Website design. *Journal of Interactive Advertising, 17*(4), 24-37.

Wilhelm, R. (2003). *Cybertrust: An economic imperative. strategy + business*, 31.

Williamson, O.E. (1994). Transaction cost economics and organization theory. In N.J. Smelser & R. Swedberg (eds.), *The handbook of economic sociology* (pp. 77-107). Princeton, NJ: Princeton University Press.

Wilson, S.G., & Abel, I. (2002). So you want to get involved in e-commerce. *Industrial Marketing Management, 31*(2), 85-94.

Wind, Y., Mahajan, V., & Gunther, J. (2002). *Convergence marketing: Strategies for reaching the new hybrid consumer*. Upper Saddle River, NJ: Prentice Hall.

Xue, M., Harker, P.T., & Heim, G.R. (2000). *Website efficiency, customer satisfaction and customer loyalty: A customer value driven perspective* (OPIM Working Paper No. 00-12-03). Wharton School, Department of Operations and Information Management.

Chapter II

Unlocking E-Customer Loyalty

Alvin Y. C. Yeo, University of Western Australia, Australia

Michael K.M. Chiam, University of Western Australia, Australia

Abstract

Marketers are working to improve loyalty apace. This chapter introduces an integrative framework for examining the relative impacts of corporate image, customer trust, and customer value on e-customer loyalty. Importantly, the authors hope that the accompanying suggested strategies would enable marketing managers to craft more compelling value propositions and effective marketing-mix strategies.

Introduction

One reason for loyalty's growing eminence is that businesses are beginning to understand the profit effect of loyal customers (Oliver, 1999; Zeithaml, Berry & Parasuraman, 1996). Marketers with loyal consumers can expect repeat patronage to remain high until competitors can find a way to close the gap in attitude among brands either by (1) trying to reduce the differential advantage of the leading brand, (2) increase the differentiation of their own brand, or (3) encourage spurious loyalty from consumers (Dick & Basu, 1994). In fact, in Reicheld's (2001) seminal book *Loyalty Rules!*, he asserts that the fundamental task of businesses today should be managing customer loyalty. Loyalty leads to higher retention. According to one study, a 5% increase in customer retention rates increases profits by 25% to 95% (Reicheld & Schefter, 2000).

It is thus heartening to note that "One of the most exciting and successful uses of [the Internet] ... may be the Internet's role in building customer loyalty and maximizing sales to your existing customers" (Griffin, 1996, p. 50). The presence of numerous Web sites means that e-retailers have a tenuous hold, at best, on a large number of "eyeballs" (Srinivasan, Anderson & Ponnavalu, 2002). Other research suggests that price sensitivity may actually be lower online than off-line (e.g., Degeratu, Rangaswamy & Wu, 2000; Lynch & Ariely, 2000). Perhaps this is because consumers can easily compare and contrast competing products and services with minimal time or effort. By changing the Internet's roles of content, context, and infrastructure, it can heighten the importance of online loyalty.

It is thus not surprising that companies are embracing the customer loyalty challenge in a variety of ways. For example, Cisco customers log in problems they are having with Cisco hardware (e.g., error rates) to a public database. This information is accessible to everyone, including media, competitors, and customers. This example suggests that accountability and open honest management practices are pivotal to loyal relationships.

Given its relative importance, it is thus surprising that relatively little has been done in conceptualizing and validating e-loyalty models (Luarn & Lin, 2003). Parasuraman and Grewal (2000) argue for more research pertaining to the influence of technology on customer responses, such as perceived value and customer loyalty. In addition, despite the fact that the impact of customer trust on customer loyalty has been well-established for online environments (e.g., Anderson & Srinivasan, 2003; Kim & Benbasat, 2003; Koehn, 2003), researchers have focused more efforts on understanding interpersonal trust and inter-organizational trust and less so on trust between people and organizations (Lee & Turban, 2001). We will examine this type of trust closer, in particular in the context of consumer e-commerce. Our study further incorporates two constructs–corporate image and customer value–that have been poorly explored in online environments despite their recognized importance in off-line contexts.

Consequently, a primary objective of this chapter is to discuss the impact of three constructs (i.e., customer trust, corporate image, and perceived value) on e-loyalty in a business-to-consumer (B2C) e-commerce context. In doing so, our model is expected to offer useful suggestions on how to manage customer trust, corporate image, and perceived value as online loyalty management tools. This chapter is generally divided into three sections. The first section will discuss the constructs of interest and clarify what they mean. In the second section, we will propose hypotheses explaining these relationships. In the final section, we introduce actionable strategies for online loyalty management based on the proposed framework.

A Brief Literature Review

Customer Loyalty

In research conducted in the 1960s and 1970s, customer loyalty was interpreted as a form of customer behavior (i.e., repeat purchasing) directed toward a particular brand over time (Tucker, 1964). However, Day (1969) criticizes behavioral conceptualizations and argues that loyalty has an attitudinal component.

More recently, Morgan and Hunt (1994) defined loyalty as "an ongoing relationship with another that is so important as to warrant maximum efforts at maintaining it" (p. 23) which implies strong affective and behavioral commitment to the company. However, this view is not universally held. For example, Assael (1987) argues that they are synonymous and represent each other.

In this chapter, we subscribe to the former school of thought where the notion of e-consumer loyalty includes both attitudinal and behavioral loyalty. This is consistent with Oliver (1999), where e-loyalty herein is defined as "A deeply held intention to repurchase a preferred product/service consistently from a particular e-vendor in the future, despite the presence of factors or circumstances that may induce switching behavior".

Customer Trust

Online markets are different from the traditional brick-and-mortar marketplaces due to the lack of face-to-face personal contact and the opportunity for buyers to see products physically (Ba, Whinston & Zhang, 1999). There are inherent risks in trading online due to information asymmetry. All these factors make trust crucial to e-commerce because it lowers transaction risks (Mayer, Davis & Schoorman, 1995). As a result, it puts pressure on online marketers to nurture stronger feelings of trust than is required in off-line environments (Keen, 1997).

Benevolence, honesty, and competence are essential components of which trust consists (Ganesan, 1994). However, trust in a networked economy is a complicated multidimensional construct (Houston et al., 2001). For example, Komiak and Benbasat (2004) differentiated between cognitive trust and emotional trust in agent-mediated e-commerce, Web-mediated e-commerce, and traditional commerce.

In this study, we adopted a parsimonious definition which is in tandem with those by Gefen, Karahanna, and Straub (2003) and Luarn and Lin (2003). E-trust is defined herein as "A set of specific beliefs dealing primarily with integrity (trustee and honesty and promise keeping), benevolence (trustee caring and motivation to act in the truster's interest), competence (ability of trustee to do what truster needs) and predictability (trustee's behavioral consistency of a particular e-vendor)".

Corporate Image

A growing number of companies have tried to position themselves through the communication channel with the objective of building strong corporate images in order to create relative attractiveness (Andreassen & Lindestad, 1998). A favorable image is a powerful tool not only for encouraging customers to choose the company's products and services but also for improving their attitudes and levels of satisfaction toward the company (Aaker, 1992, p. 16).

A review of the literature, however, reveals scant research on the concept of corporate image in online environments. Much of the traditional research focuses on products (e.g., Darden & Schwinghammer, 1985; Render & O'Connor, 1976) or services (Gronroos, 1990) in off-line environments. There is an urgent need for more of such research that explores this concept better as it applies to online environments. Customers are often overwhelmed with a variety of offerings on the Internet. As a result, they base their decisions on global judgments, such as store image and reputation (Teas & Agarwal, 2000).

Generally, image has been treated as a gestalt, reflecting a customer's overall impression (Dichter, 1985; Zimmer & Golden, 1988) or as an "idiosyncratic cognitive configuration" (Mazursky & Jacoby, 1986). Also, Kennedy (1977) distinguished between functional image and emotional image. More recent researchers have adopted broader and richer definitions that emphasize the consumption experience of consumers. For example, Andreassen & Lindestad (1998) defined corporate image as "A function of the accumulation of purchasing/consumption experience over time" (p. 84).

In consonance with definitions by Barich and Kotler (1991) and Kotler (1991), corporate e-image is thus defined in this chapter as "an overall impression held of an e-vendor by its customers at a particular point in time. This, in turn, is the net result of consumers' experiences with an organization, both online and offline, and from the processing of information on the attributes that constitute functional indicators of image".

Customer Value

The inclusion of the value construct in our model is important for several reasons. First, recent research revealed that "More than 70% of customers feel they gain nothing by being loyal to a company" (Gillespie, 1999, p. 2) and, therefore, would move if perceived benefits were greater elsewhere. This makes understanding what buyers value within a given offering, creating value for them, and then managing it over time, essential elements of every market-oriented firm's core business strategy (DeSarbo, Jedidi & Sinha, 2001; Slater & Narver, 1998).

Second, Helm and Sinha (2001) argue for the importance of delivering customer value in electronic B2C operations. However, despite the increased attention being given to customer value (Band, 1991; Gale, 1994), there has been little empirical research to develop an in-depth understanding of the concept (Holbrook, 1994; Sweeney & Soutar, 2001). Fewer studies still (e.g.,

Chen & Dubinsky, 2003) have examined this construct in the context of online environments. Finally, to the best of our knowledge, there seems to be no research that has captured the relationships between corporate image, customer trust, and perceived value in a single integrative framework.

Perceived value has been defined generally as "a trade-off of the salient give and get components" (Zeithaml, 1988, p. 14). However, several researchers have noted that perceived value is a more complex construct, in which notions such as perceived price, quality, benefits, and sacrifice are embedded (Bolton & Drew, 1991; Holbrook, 1994). Based on a synthesis of previous definitions (Chen & Dubinsky, 2003; Woodruff, 1997), perceived customer value is defined here as "A consumer's perception of the benefits gained in exchange for the costs incurred to attain their goals at a particular point in time".

Proposed Model for Business-to-Consumer E-Commerce

Consumers rely heavily on the online vendor's image as a proxy for trustworthiness (Jarvenpaa & Tractinsky, 1999; Lee & Turban, 2001). This is because of the lack of intrinsic product cues that are generally used to evaluate quality. Indeed, Yoon (2002) found that Web site trust shown is significantly related to corporate awareness and image. Specifically, Einwiller (2003) found that a reputation-trust relationship is particularly pronounced if the consumer's experience with online shopping is low. Customers who had gained much experience with a particular retailer were significantly less influenced by retailer's reputation than those who had never or rarely bought something from the respective retailer. We thus posit that *corporate image is positively related to customer trust (H₁)*.

Chen and Dubinsky (2003) reported that online retailer image is a forward indicator of value in the online context. This is corroborated by Strader and Shaw (1999) who found that in Internet marketing, unless a seller's price is significantly lower than prices of a trusted seller, switching costs will inhibit the consumer from buying from the unknown e-seller. Thus, we argue that *corporate image is positively related to perceived value (H₂)*.

Meanwhile, Sirdeshmukh, Singh, and Sabol (2002) suggested that perceived value is an important partial mediator of the trust-loyalty relationship. Customer relationship value may only develop when the customer has "confidence in an exchange partner's reliability and integrity" (Morgan & Hunt, 1994, p. 23). Safety, credibility, and security are important to reduce the sacrifice for the customer in a relationship and therefore lead to higher e-relationship value. Hence, we postulate that *customer trust is positively related to perceived value (H₃)*.

Finally, Hoffman and Novak (1996) argued that the likelihood of Internet product purchase is influenced by the amount of consumer trust regarding the delivery of goods and use of personal information. Trust influences loyalty by affecting the consumer's perception of congruence in values with the good/service provider (Gwinner, Gremler & Bitner, 1998). Also, when there is perceived similarity in values between the firm and the consumer, the consumer is more entrenched in a relationship. This is also evidenced in research by Gefen (2000) and Chircu, Davis, and Kaufman (2000) which showed that customer trust has a positive effect on e-adoption intention. We thus propose that *customer trust is positively related to e-customer loyalty (H₄)* and *perceived value is positively related to e-customer loyalty (H₅)*. These relationships are captured in a conceptual model in Figure 1.

Figure 1. An integrative framework for understanding e-customer loyalty

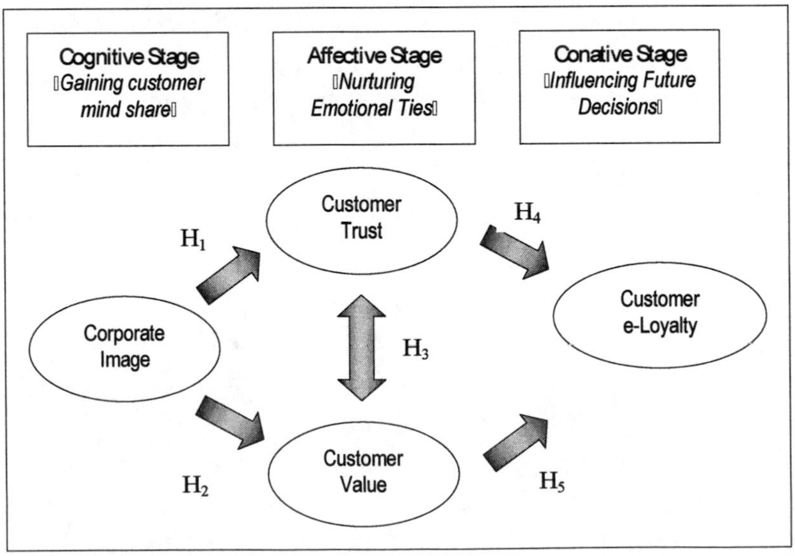

Realizing E-Customer Loyalty: Managerial Recommendations

This next section suggests three strategies to attain the *tao of loyalty* through effective management of each of the three constructs of interest: building trust, realizing corporate image equity, and providing customer value.

Build Trust

As discussed earlier, lack of trust is one of the main reasons that consumers and companies do not engage in e-commerce (Castelfranchi & Tan, 2001). We argue that efforts to build trust should focus on three areas – transactional security trust, information privacy trust, and exchange trust – each with its own function in motivating consumer participation in exchange.

Transactional security trust refers to, first, the customers' belief about the e-tailer's expertise in providing a secure shopping environment and, second, the consumer's expectation about an e-tailer's ability to protect personal information from unauthorized access by third parties (i.e., hackers). Typically, customers expect that an online company's Web site will provide methods for secure exchange of financial information with them (Ranganathan & Ganapathy, 2002). To assuage customers on this aspect, firms should put in place reliable control systems that can be easily understood by customers. Expedia.com, an online travel agency, explains to its customers that when personal information is sent over the Internet, their data is protected by Secure Socket Layer (SSL) technology to ensure safe transmission. To further assuage concerns about the robustness of its security systems, the firm also obtained certification from RSA Data Security, Inc., a top maker of security-related hardware and software.

Like transactional security trust, *information privacy trust* is conceptualized in terms of two dimensions – unauthorized tracking and unauthorized information dissemination – that refer to an e-tailer's privacy protection responsibilities (Hoffman & Novak, 1996). Research has shown that fairness of a company's Web site with respect to information privacy is a significant factor in building trust and in ensuring the continuation of the relationship with the e-retailer (Culnan & Armstrong, 1999). For example, Expedia, an online travel portal, has an extensive privacy policy detailing the type of information that is collated, with whom the information is shared, and the control users have over their information. An audit firm, PricewaterhouseCoopers LLP, is hired to ensure compliance in their information privacy practices.

Exchange trust is defined in terms of competence and benevolence as they apply to an e-tailer's promise fulfillment function (Sirdeshmukh, Singh & Sabol, 2002). Specifically, exchange trust is customer confidence that the e-tailer (1) will fulfill its transaction-specific obligations consistent with the terms of the purchase agreement or other internally-held reference standards developed as a result of interaction with other e-tailers or nonstore retailers (competence) and (2) will not engage in opportunistic behavior at any time in the process of product or service delivery (benevolence). A type of service, online virtual community, is emerging to mitigate exchange risks. One of the better-known ones, eBay's Feedback Forum, provides information on sellers' reputation based on feedback from previous trades. Traders having high reputations may enjoy price premiums in online market competitions.

It should be highlighted that trust in the e-tailer does not have to be a necessary condition to purchase online. It has been argued that lack of trust in organizations can be offset by trust in the control system (Tan & Thoen, 2000). In this respect, a trusted third party model appears also to be taking shape. For example, eBay imparts to buyers and sellers the confidence to trade online by providing financial instruments, such as Cyber Cash and escrow, establishing guarantees for transactions and addressing privacy and delivery reliability concerns. Finally, a wide variety of so-called trust seals (e.g., Trust-e, WebTrust, BBB online) have been introduced for use on Web sites. These seals are issued by organizations that certify the procedural controls of Web sites (e.g., that the Web site handles payments securely and reliably).

Realize Corporate Image Equity

Research has shown that a Web presence complements traditional brick-and-mortar stores by helping to create awareness, make impressions, supply details, drive store traffic, and generate prospects (Maddox, Mehta & Daubek, 1997). Having a Web site is important for brand building and creating a favorable corporate image (Berthon, Pitt & Watson, 1996; Farr, 1999). However, that alone is not sufficient.

First, firms should strive to create a strong Internet identity through a distinctive personality. Recognizing that a company's name identifies it to outside observers (Aaker, 1991), For example, Tiffany, the well-known jewelry retailer, invested substantially in digital imaging technology to ensure that all images of jewelry on its Web site are presented using high quality graphics. The overall impact of the Web site reinforces Tiffany's reputation as a prestigious high-quality retailer.

Second, firms should ensure there is sustained buzz about the firm's presence. Many online firms use banner advertisements, pop-up windows, and mass e-mails to promote their products. Advertising directly confers brand and product information to consumers, and it serves to increase consumers' awareness and knowledge of a particular brand (Vakratsas & Ambler, 1999). High levels of advertising expenditures send a strong signal about a firm's commitment to quality (Shapiro, 1983). For example, Amazon.com invests more than 23% of its revenues in advertising, thus making it one of the most well-known companies in online book retail business.

While advertising remains an important element of the online business model, firms should strive to have a balanced mix of online advertising and off-line brand communications. One important off-line communication source that companies should pay more attention to is interpersonal encounters or word-of-mouth (WOM). According to Buttle (1998), WOM is, in general, more influential on behavior than other organization-controlled sources.

Finally, firms should strive to earn a reputation for being innovative. Generally, research has found that early movers enjoy initial and sustain continuous market share advantages. For example, Robinson and Fornell (1985) suggested that being a pioneer or product leader gives consumers a favorable image and higher familiarity on the firm's products. Examples of online firms that attain such leadership include Priceline.com, which pioneered the "name your own price" practice in its travel booking service, and eBay, which introduced the proxy bidding technology and feedback mechanism in its Internet auction business.

Create Value Experiences Online

In this section, actionable strategies relating to two important areas – target marketing and enriching the customers' online experience – will be discussed.

Target Marketing. A thorough understanding of one's target audience is critical. With a better understanding of the target market, businesses will be able to better design their Web sites to match preferences of their target groups. Two crucial areas need to be addressed: How do customers make purchasing decisions, and what information do they need in making those decisions? Wine.com, a leading purveyor of wine and gourmet products, originally believed that its core customers were wine connoisseurs. A careful assessment of online data revealed that its best customers were wine novices. With this knowledge, Wine.com significantly modified its Web site to deliver the features these less experienced consumers wanted–recommendations and educational content.

Enrich the Customers' Online Experience. The natural question, once marketing managers know who their best customers are, is how to enrich these customers' overall experience. One suggestion marketing managers may want to focus their efforts on is customization. Customization increases the probability that customers will find something that they wish to buy, creates the perception of increased choice by enabling a quick focus on what the customer really wants (Shostak, 1987), signals high quality (Ostrom & Iacobucci, 1995), and drives customers to use simplistic decision rules to narrow down the alternatives (Kahn, 1998). In this respect, customization has often been regarded as one of the essential determinants of customer loyalty (Reynolds & Beatty, 1999).

Managers may also want to focus on improving Web site convenience. A convenient Web site provides a short response time, facilitates fast completion of a transaction, and minimizes customer effort. For example, Weyerhauser, a forest products company, manages its door-making through a customized Internet system in which the customer inputs specifications that are then electronically digitized into profile information that drives the manufacture of that particular door. By providing convenience and customized experience for customers, Weyerhauser is able to enhance its customers' online experience.

Conclusions

Marketers are perpetually seeking to increase customer loyalty, which perhaps is the only source of sustainable competitive advantage in the coming decade for many firms. This chapter suggests strategies for firms to develop an e-customer relationship orientation and improve e-customer loyalty. In this aspect, corporate image, customer trust, and customer value were found to be important drivers of e-customer loyalty.

To enable more compelling value propositions to be crafted, future research could examine the moderating effects of consumer attributes like involvement and affect on the model. For example, depending on the level of involvement, consumers may be passive or active when receiving advertising communication, which in turn could moderate their perceptions of value and trust levels. Empirical comparison between different cultures could also be collated for greater generalizations of the proposed model.

Acknowledgments

The authors gratefully acknowledge the valuable comments and suggestions by the anonymous reviewers, Theresa B. Flaherty, and Irvine Clarke III.

References

Aaker, D. A. (1991). *Managing brand equity: Capitalising on the value of a brand name*. New York: The Free Press.

Aaker, D. (1992). *Strategic market management*. New York: John Wiley & Sons.

Ajzen, I., & Fishbein, M. (1980). *Understanding attitudes and predicting social behaviour*. Englewoods Cliffs, NJ: Prentice Hall.

Anderson, R.E., & Srinivasan, S.S. (2003). E-satisfaction and e-loyalty: A contingency approach. *Psychology & Marketing, 20*(2), 123-132.

Andreassen, T.W., & Lindestad, B. (1998). The effects of corporate image in the formation of customer loyalty. *Journal of Service Marketing, 1,* 82-92.

Assael, H. (1987). *Consumer behavior and marketing action*. Boston: Kent Publishing.

Ba, S.A., Whinston, A., & Zhang, H. (1999, December 13-15). *Building trust in the electronic market through an economic incentive mechanism*. Paper presented at the 20th International Conference on Information Systems, Charlotte, NC.

Band, W. (1991). *Creating value for customers*. New York: John Wiley & Sons.

Barich, H., & Kotler, P. (1991). A framework for marketing image management. *Sloan Management Review,* Winter, 94-104.

Berthon, P., Pitt, L., & Watson, R. (1996). The World Wide Web as an advertising medium: Towards an understanding of conversion efficiency. *Journal of Advertising Research, 60*(1), 43-54.

Bolton, R.N., & Drew, J.H. (1991, March). A multistage model of customers' assessments of service quality and value. *Journal of Consumer Research, 17*, 375-384.

Buttle, F.A. (1998). Word of mouth: Understanding and managing referral marketing. *Journal of Strategic Marketing, 6*, 241-254.

Castelfranchi, C., & Tan, Y.H. (2001, January 3-6). *The role of trust and deception in virtual societies.* Paper presented at the Hawaii International Conference on System Sciences, Maui, Hawaii.

Chen, Z., & Dubinsky, A.J. (2003). A conceptual model of perceived customer value in e-commerce: A preliminary investigation. *Psychology & Marketing, 204*, 323-347.

Chircu, A.M., Davis, G.B., & Kaufman, R.J. (2000, August 10-13). *Trust, expertise and e-commerce intermediary adoption.* Paper presented at the 6th Americas Conference on Information Systems, Long Beach, CA.

Culnan, M.J., & Armstrong, P.K. (1999). Information privacy concerns, procedural fairness, and impersonal trust: An empirical investigation. *Organization Science, 10*(1), 104-115.

Darden, W., & Schwinghammer, J. (1985). The influence of social characteristics on perceived quality. In J. Jacoby & J. Olson (Eds.), *Patronage choice behaviour and perceived quality: How consumers view new stores and merchandise* (pp. 161-172). Lexington, MA: DC Health.

Day, G. (1969). A two-dimensional concept of brand loyalty. *Journal of Advertising Research, 9*, 29-35.

Degeratu, A., Rangaswamy, A., & Wu, J. (2000). Consumer choice behaviour in online and traditional supermarkets: The effects of brand name, price and other search attributes. *International Journal of Research In Marketing, 17*(1), 55-78.

DeSarbo, W.S., Jedidi, S., & Sinha, I. (2001). An empirical investigation of the structural antecedents of perceived customer value in a heterogeneous population. *Strategic Management Journal, 22*(9), 455-468.

Dichter, E. (1985). What's in an image. *Journal of Consumer Marketing, 2*(1), 75-81.

Dick, A.S., & Basu, K. (1994). Costomer loyalty: Toward an integrated conceptual framework. *Journal of the Acadamy of Marketing Science, 22*, 99 -113.

Einwiller, S. (2003). When reputation engenders trust: An empirical investigation in business-to-consumer electronic commerce. *Electronic Markets, 13*(3), 196-209.

Farr, A. (1999). How advertisers build brand equity. In H. Jones (ed.), *How to use advertising to build strong brands* (pp. 000-000). Thousand Oaks, CA: Sage.

Gale, B.T. (1994). *Managing customer value: Creating quality and service that customers can see.* New York: Free Press.

Ganesan, S. (1994, April). Determinants of long-term orientation in buyer-seller relationships. *Journal of Marketing, 58*, 1-19.

Gefen, D. (2000). E-commerce: The role of familiarity and trust. *Omega: The International Journal of Management Science, 28*(5), 725-737.

Gefen, D., Karahanna, E., & Straub, D.W. (2003). Trust and TAM in online shopping: An integrated model. *MIS Quarterly, 27*(1), 51-90.

Gillespie, J. (1999). Surprising news about loyalty. *Communication Briefings, 19*, 1-8.

Griffin, J. (1996). The Internet's expanding role in building customer loyalty. *Direct Marketing, 59*(7), 50-53.

Gronroos, C. (1990). *Service management and marketing: Managing the moments of truth in service competition.* Lexington, MA: Lexington Books.

Gwinner, K.P., Gremler, D.D., & Bitner, M.J. (1998). Relational benefits in service industries: The customer's perspective. *Journal of the Academy of the Marketing Science, 26*(Spring), 101-114.

Helm, G.R., & Sinha, K.K. (2001). A product-process matrix for electronic B2C operations: Implications for the delivery of customer value. *Journal of Service Research, 3*(4), 286-299.

Hoffman, D.L., & Novak, T.O. (1996, July). Marketing in hypermedia computer-mediated environments: Conceptual foundations. *Journal of Marketing, 60*, 50-68.

Holbrook, M.B. (1994). The nature of customer value: An axiology of services in the consumption experience. In R.L. Oliver & R.T. Rust (Eds.), *Service quality: New directions in theory and practice*. Thousand Oaks, CA: Sage.

Houston, M.B., Walker, B.A., Hutt, M.D., & Reingen, P. (2001). Cross-unit competition for a market charter: The enduring influence of structure. *Journal of Marketing, 65*(2), 19-34.

Jarvenpaa, S.L., & Tractinsky, N. (1999). Consumer trust in an Internet store: A cross-cultural validation. *Journal of Computer-Mediated Communication [On-line], 5* (1).

Kahn, B. (1998). Dynamic relationships with customers: High-variety strategies. *Journal of Academy of Marketing Science, 26*(Winter), 45-53.

Keen, P.G.W. (1997, April). Are you ready for "trust" economy? *Computer World, 31*(16), 80.

Kennedy, S.H. (1977). Nurturing corporate image. *European Journal of Marketing, 11*(3), 120-164.

Kim, D., & Benbasat, I. (2003). Trust-related arguments in Internet stores: A framework for evaluation. *Journal of Electronic Commerce Research, 4*(2), 49-64.

Koehn, D. (2003). The nature of and conditions for online trust. *Journal of Business Ethics, 43*(1/2), 3-19.

Komiak, S.X., & Benbasat, I. (2004). Understanding customer trust in agent-mediated electronic commerce, Web-mediated electronic commerce, and traditional commerce. *Information Technology & Management, 5*(1/2), 181.

Kotler, P. (1991). *Marketing management*. Englewood Cliffs, NJ: Prentice Hall.

Lee, M.K.O., & Turban, E. (2001). A trust model for consumer Internet shopping. *International Journal of Electronic Commerce, 6*(4), 75-91.

Luarn, P., & Lin, H.H. (2003). A customer loyalty model for e-service context. *Journal of Electronic Commerce Research, 4*(4), 156-167.

Lynch Jr., J., & Ariely, D. (2000). Wine online: Search costs affect competition on price, quality and dstribution. *Marketing Science, 19*(1), 83-104.

Maddox, L., Mehta, D., & Daubek, H. (1997). The role and effect of Web addresses in advertising. *Journal of Advertising Research, 37*(2), 47-59.

Mayer, R.C., Davis, J.H., & Schoorman, D.F. (1995). An integrative model of organization trust. *Academy of Management Review, 20*(3), 709-734.

Mazursky, D., & Jacoby, J. (1986). Exploring the development of store images. *Journal of Retailing, 62*(2), 145-165.

Morgan, R.M., & Hunt, S. (1994). The commitment-trust theory of relationship marketing. *Journal of Marketing, 58*(July), 20-38.

Oliver, R.L. (1999). Whence customer loyalty? *Journal of Marketing, 63*(Special Issue), 33-44.

Ostrom, A., & Iacobucci, D. (1995, January). Consumer tradeoffs and evaluation of services. *Journal of Marketing, 59*, 17-28.

Parasuraman, A., & Grewal, D. (2000). Serving customers and consumers effectively in the twentieth century: A conceptual framework and overview. *Journal of Academy of Marketing Science, 28*(1), 9-16.

Ranganathan, C., & Ganapathy, S. (2002). Key dimensions of business-to-consumer Web sites. *Information & Management, 39*(6), 457.

Reicheld, F.F. (2001). *Loyalty rules: How today's leaders build lasting relationships* (2nd ed.). MA, US: Harvard Business School Press.

Reicheld, F.F., & Schefter, P. (2000, July-August). E-loyalty: Your secret weapon on the Web. *Harvard Business Review, 78*, 105-113.

Render, B., & O'Connor, T. (1976). The influence of price, store name and brand name on perception of product quality. *Journal of the Academy of Marketing Science, 4*(Fall), 772-730.

Reynolds, K.E., & Beatty, S.E. (1999). Customer benefits and company consequences of customer-salesperson relationships in retailing. *Journal of Retailing, 75*(1), 11-32.

Robinson, W., & Fornell, C. (1985, August). Sources of market pioneer advantages in consumer goods industries. *Journal of Marketing Research, 22*, 305-317.

Shapiro, C. (1983). Premiums for high-quality products as returns to reputations. *Quarterly Journal of Economics, 98*, 659-681.

Shostak, G.L. (1987, April). Breaking free from product marketing. *Journal of Marketing, 41*, 73-80.

Sirdeshmukh, D., Singh, J., & Sabol, B. (2002, January). Consumer trust, value and loyalty in relational exchanges. *Journal of Marketing, 66*, 15-37.

Slater, S., & Narver, J. (1998). Customer-led and market-oriented: Let's not confuse the two. *Strategic Management Journal, 19*(10), 1001-1006.

Srinivasan, S.S., Anderson, R., & Ponnavalu, K. (2002). Customer loyalty in e-commerce: An exploration of its antecedents and consequences. *Journal of Retailing, 78*, 41-50.

Strader, T.J., & Shaw, M.J. (1999). Consumer cost differences for traditional and Internet markets. *Internet Research, 9*(2), 82-92.

Sweeney, J.C., & Soutar, G.N. (2001). Consumer perceived value: The development of a multiple item scale. *Journal of Retailing, 77*, 203-220.

Tan, Y.H., & Thoen, W. (2000). Toward a generic model of trust for electronic commerce. *International Journal of Electronic Commerce, 5*(2), 61-74.

Teas, R.K., & Agarwal, S. (2000). The effect of extrinsic product cues on consumers' perceptions of quality, sacrifice, and value. *Journal of the Academy of Marketing Science, 28*(2), 278-290.

Tucker, W.T. (1964). The development of brand loyalty. *Journal of Marketing Research, 1*, 32-35.

Vakratsas, D., & Ambler, T. (1999). How advertising works: What do we really know? *Journal of Marketing, 36*, 26-43.

Woodruff, R.B. (1997). Customer value: The next source for competitive advantage. *Journal of Academy of Marketing Science, 25*(2), 139-153.

Yoon, S.J. (2002). The antecedents and consequences of trust in online-purchase decisions. *Journal of Interactive Marketing, 16*(2), 47-63.

Zeithaml, V.A. (1988, July). Consumer perceptions of price, quality and value: A means-end model and synthesis of evidence. *Journal of Marketing, 52*, 2-22.

Zeithaml, V.A., Berry, L.L., & Parasuraman, A. (1996, April). The behavioral consequences of service quality. *Journal of Marketing, 60*, 31-46.

Zimmer, M.R., & Golden, L.L. (1988). Impression of retail stores: A content analysis of consumer images. *Journal of Retailing, 64*(3), 265-293.

Chapter III

Drivers and Barriers to Online Shopping:
The Interaction of Product, Consumer, and Retailer Factors

Francesca Dall'Olmo Riley, Kingston University, UK

Daniele Scarpi, Universita' di Bologna, Italy

Angelo Manaresi, Universita' di Bologna, Italy

Abstract

Through a review of the literature, this chapter focuses on three key influences on purchase behavior on the Internet: product, consumer, and retailer factors. Product characteristics and branding not only influence many consumer-related factors (e.g., the need to handle the product and risk perceptions), but also affect retailers' strategic and tactical online decisions (e.g., the balance between off-line and online retail provision and the breadth and depth of products and brands selection). This chapter also examines how consumer-related factors (e.g., consumers' expertise, attitudes toward the Internet, and shopping orientation) affect online purchasing and the implications for e-retailers. Finally, the chapter discusses how e-retailers marketing efforts (retailer factors) can be used to overcome the barriers to Web purchasing resulting from specific product and consumer related characteristics. Clear and easy to implement recommendations to managers are offered.

Introduction

Although the statistics change almost on a daily basis, the use of the Web as a distribution channel remains limited, in spite of the ever-increasing use of the Internet for other purposes. Internet sales figures indicate that online selling is far from replacing traditional channels or other nonstore retailers. For instance, online expenditure in the UK amount to about 6% of total retail spending (Interactive Media in Retail Group as cited in Hall, 2003). Many Internet users appear reluctant to shop online and use the Internet only as a means of gathering information before purchasing in a traditional brick-and-mortar environment. Dieringer Research Group (cited in Mazur, 2003) reports that in the past year $138 billion were spent by U.S. consumers for purchasing products off-line after seeking information online, compared with $95 billion spent for shopping directly online. Finally, many brick-and-mortar retailers do not have a presence on the Web or, if they do, they use their sites as a communication and promotional tool rather than for selling (compare Hart, Doherty & Ellis-Chadwick, 2000).

Retailers and other organizations that either already have an online selling facility or are considering doing so would benefit from a better understanding of the factors that influence (either positively or negatively) consumers' motivations to purchase online. This understanding would help managers to better plan their Internet strategies, design their Web sites more effectively, select the assortment of goods more likely to sell online, and attract a larger number of Internet buyers, converting browsers into shoppers.

The first objective of this chapter is to provide a synthesis of the key factors that influence consumer purchasing behavior on the Internet, starting with (1) product-related factors (and the effect of brand name) and (2) consumer-related factors. Although the literature considers these two factors from many different perspectives, there is shared agreement about their importance for understanding (and directing) consumers' behavior online. A review of these factors not only will be useful to academic researchers but may also help managers to improve the effectiveness of their online strategies. Appropriate retailer factors (i.e., retailers' Internet strategies and tactics) are further key elements influencing how the Internet is used (e.g., for buying rather than browsing) and how the overall online shopping experience is evaluated (compare Cowles, Kiecker & Little, 2002). Conversely, while new technologies can enhance the shopping experience, their application "must be tailored to the unique requirements of consumer segments and product categories" (Burke, 2002, p. 411). The second objective of this chapter is therefore to provide practical suggestions for retailers' Internet strategies and tactics (retailer factors).

Building on the model proposed by Cowles et al. (2002), this chapter is structured on the framework depicted in Figure 1. First, we review the literature concerning product-related factors: product typologies and the effect of brand name within an Internet mediated environment. Product characteristics are

Figure 1. The interaction of product, consumer and retailer factors

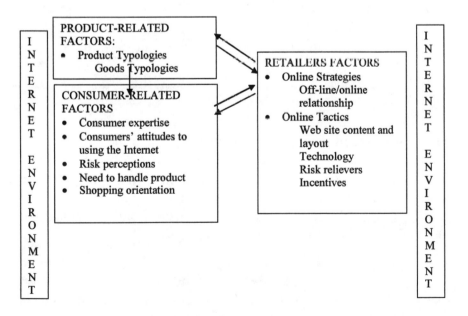

found not only to influence many consumer-related factors (e.g., the need to handle the product and risk perceptions) but also to affect retailers' strategic and tactical online decisions (e.g., the relationship between off-line and online provision and the Web site content and layout). We then go on to analyze consumer-related factors (e.g., consumer expertise, attitudes toward the Internet, and shopping orientation), how they affect online purchasing, and the implications for e-retailers. Finally, we discuss how e-retailers marketing efforts can be used to overcome the barriers to Web purchasing resulting from specific product and consumer related characteristics. We aim to offer clear and easy to implement recommendations to managers.

Product Related Factors

Researchers agree that product type and related characteristics have a significant effect not only on consumers' online purchase intentions and behavior (Brown, Pope & Voges, 2003; Peterson, Balasubramanian & Bronnenberg, 1997) but also on e-retail decisions, including Web site design and assortment planning (Alba et al., 1997; Burke, 1997, 2002). In this section, first we review general product typologies, then we analyze specific goods and services typologies, their impact on consumers' online purchase intentions and behavior, and relate managerial implications. Finally, we discuss issues concerning the importance of branding.

Product Typologies

According to Peterson et al. (1997), some products are more suitable than others to Internet selling because of the characteristics they possess: when the core essence of the product is of an intangible and informative nature, the Internet is best suited as a distribution channel. Consistently, Brown et al. (2003) found product type to have a significant impact on online purchase intentions, especially when considered in conjunction with individual shopping preferences and prior online purchase experience.

Klein (1998) categorizes products on the basis of their *information content*, specifically the balance between *search* and *experience* attributes they possess. This categorization is based on Nelson's (1974) distinction between search and experience products. Search products are those for which full information on the most important attributes can be obtained prior to purchase. In contrast, for experience products, *either* full information on dominant attributes cannot be attained without direct experience *or* information search is more costly/difficult than direct product experience. Many service products (e.g., a package holiday) are lower in search than in experience attributes, (i.e., their quality cannot be fully evaluated until purchase and consumption is taking place).

By facilitating the acquisition and accessibility of information and by lowering the cost of searching, the Internet might not only suit search products but may also enable the transformation of experience products into search products. For instance, a *virtual experience* may be provided by means of online demonstrations and third-party endorsement. Such virtual experience may also succeed in lowering consumers' perceived risk (Klein, 1998).

On the other hand, on the online medium, the risk that a brand or product will not perform as expected (product performance risk) often originates from the impossibility to touch, feel, and try the good or from insufficient information on quality attributes relevant to the consumer (Forsythe & Shi, 2003). Some products, such as clothes, whose quality attributes (fit, quality of material, feel, etc.) can be assessed before purchase when bought in a traditional brick-and-mortar retail environment, shift toward becoming experience products when sold online. When shopping for clothes on the Internet, consumers often have to rely either on other attributes, such as the reputation of the brand or on experience acquired off-line (e.g., from visiting a retail outlet).

Therefore, the characteristics of the good/service have to be considered in conjunction with the peculiarities of the online distribution medium if we want to gain a deeper insight and derive implications for practice (compare Alba et al., 1997). This is the approach that we will take in the following sections through the examination of (1) goods typology; (2) services typology; and (3) the effect of brand names.

Goods Typology

Burke (1997) suggests that the Internet is an unsuitable distribution channel for heavy, bulky, and fragile items; low-margin items; products requiring in-store demonstration; and products that are needed urgently. In contrast, when the delivery cost accounts for a small proportion of the cost of the good, because the good is expensive and infrequently purchased, Internet selling becomes possible (Peterson et al., 1997). Indeed, some expensive and infrequently purchased goods, such as personal computers, are successfully sold online (e.g., *Dell.com* and *Gateway.com*).

On the other hand, consumers' propensity to purchase a good via the Internet decreases as one moves from lower priced items to higher priced ones since the risk associated with making the wrong purchase increases (Van den Poel & Leunis, 1999). Accordingly, Lee and Tan (2003) postulate that consumers are more likely to shop online for goods that are low in purchase risk. Increased cost and the associated risk might make it necessary to inspect the good personally before purchasing online (Citrin et al., 2003; Peterson et al., 1997).

As previously noted, the need to inspect a good prior to purchase is a frequently cited barrier toward Web retailing (e.g., Fenech & O'Cass, 2001). Hence, it has been suggested that nonstandardized goods, such as clothes, requiring a high degree of sensory input before reaching a purchase decision, are less likely to be acquired online (e.g., Citrin et al., 2003). However, the need of inspection prior to purchase seems to affect only the frequency of purchasing online but not any other Internet shopping behavior (Forsythe & Shi, 2003). Furthermore, experience with the medium may act as a moderating factor (Fenech & O'Cass, 2001). Nonetheless, standardized search goods, such as books, CDs, and software, not only score high in consumers' declared online purchase preferences (e.g., Girard, Silverblatt & Korgaonkar, 2002) but also top most of the online sales statistics (e.g., Lee & Johnson, 2002).

These findings, though partially contradictory, have practical implications in terms of the assortment of goods that are offered online (e.g., clothes versus accessories) and the design of the Web site itself. For instance, the use of interactive features, which would allow potential customers to *virtually try on* an item of clothing, may be helpful in overcoming the lack of tactile or other sensory input. Furthermore, risk relievers, such as money back guarantees and a price reduction (Van den Poel & Leunis, 1999) or clear explanations of how to return goods that fail to meet expectations, should feature prominently on Web sites.

Services Typology

Peterson et al. (1997) propose that, for products of an intangible and informational nature, acquisition can easily occur on the Internet whether the potential for differentiation is high or low. The online medium should not alter significantly the core essence of services purchasing since by definition services are rich in experience and credence attributes and are therefore hard to evaluate prior to purchase and use, no matter the purchase medium. Indeed, since services do not require a *tactile input* prior to buying, consumers might be expected to find it easier to purchase services rather than tangible goods online.

However, the specific characteristics of a service offering may have a bearing on the suitability of the Internet as a delivery channel. For instance, Bowen (1990) categorized services into three groups on the basis of degree of contact with employees, customization, and personalization. High contact customized personal services, such as legal and counseling services, may be very hard to deliver online since customers perceive face-to-face contact with employees as a fundamental aspect of the service. On the other hand, for moderate contact semi-customized personal services, such as banking, the Internet may be considered by many customers as a suitable and attractive medium, providing advantages, such as time saving and the opportunity to access personal information at any time. Furthermore, moderate contact standardized services, like the booking of airline tickets, appear perfectly suited to online delivery since for these kind of services, customers value speed, consistency, and price saving the most (Lovelock, 1984).

Finally, as Alba et al. (1997) note, "a brand is a search attribute that assures consumers of a consistent level of product quality. It might be the only attribute available to assess some credence goods" (p. 49). Brands may have an important role in shifting consumers' perceptions away from experience attributes, such as tactility to search attributes such as reputation (Citrin et al., 2003), and well-known brands have been found to be important risk relievers for online shoppers (Van den Poel & Leunis, 1999). The effect of brand names as risk relievers and facilitators of online purchases for goods and services is addressed below.

Effect of Brand Name

Most researchers agree that consumers are more likely to shop online for goods with well-known brands and are more likely to shop from well-known retailers, even if they carry lesser-known brands (Kau, Tang & Ghose, 2003; Lee & Tan, 2003). In the words of Pricewaterhouse Coopers, "80% of consumers who have shopped for clothing online over the past six months do so at sites operated by a traditional store or catalogue retailer, and one-third of online consumers say they shopped for clothing at sites operated by a manufacturer whose products they were already familiar with" (Kau et al., 2003, p. 141).

Brand name, price, advertising, and packaging are important extrinsic cues facilitating consumers' choices whenever it is difficult or costly to collect and examine intrinsic product attributes, such as flavor, color, size, and texture (Klein, 1998). Hence, for all goods and services that either cannot be properly evaluated prior to online purchase or involve a high degree of purchase risk, brands fulfill an important role.

Degeratu, Rangaswamy, and Wu (2000) postulate that brand equity could have a higher impact online than off-line, as confirmed by other researchers. For instance, Balabanis and Reynolds (2001) noted a significant influence of existing brand attitudes on the attitudes of online shoppers, and Harvin (2000) found consumers to be more comfortable with companies that have strong off-line brands they are already familiar with and whom they trust. Managers should therefore invest in building the reputation of their brands to enable the reduction of consumers' perceived risk. Well-established traditional retailers will have an advantage over start-ups in electronic retailing since they can capitalize on their reputation as a risk reliever, reducing the risk aversion of online consumers (Lee & Tan, 2003). These considerations lead us to examine consumer-related factors and characteristics as possible facilitators or barriers to e-commerce.

Consumer Related Factors

A large body of literature deals with individual attitudes, characteristics, and other consumer-related factors that may influence either positively or negatively the proclivity to purchase online. An understanding of these factors will enable managers to tailor their offerings and e-marketing efforts more effectively to different consumer segments. Important differentiating factors among consumers include (1) consumer expertise; (2) consumers' attitudes to using the Internet; (3) risk perceptions; (4) need to handle products before purchase; and (5) shopping orientation. These are addressed in turn in the following sections, following a broad order of importance. It should be noted, however, that the extent to which each factor is dominant in determining (or hampering) Internet purchase probability is often dependent on the interaction with other factors, particularly the type of product being bought (Figure 1). In contrast to common belief, price sensitivity or significance of price has not been found to be a strong discriminator between Internet shoppers and nonshoppers (compare Fenech & O'Cass, 2001) and will therefore not be discussed here.

Consumer Expertise

The level of expertise with the Internet medium itself, both as an information gathering and as a shopping vehicle, is considered by many researchers as the key predictor of online buying (Bellman, Lohse & Johnson, 1999; Brown et al., 2003; Goldsmith & Bridges, 2000). Researchers agree that expertise has a positive influence on the frequency and number of online purchases, no matter what the product is (e.g., Fenech, 2000; Goldsmith & Goldsmith, 2002; Van den Poel & Leunis, 1999). Heavy users of the Internet evaluate it more favorably than light users (Van den Poel & Leunis, 1999). Expert online buyers have more positive attitudes toward the Internet in general (Goldsmith & Goldsmith, 2002) and feel less anxious toward it than novices (Fenech, 2000). Consumers' attitudes to using the Internet are discussed next.

Consumers' Attitudes to Using the Internet

Attitudes toward using technology and Internet technology, in particular, influence the propensity to shop online and the perceived usefulness of online shopping (Davis & Venkatesh, 1996). Goldsmith and Bridges (2000) classify attitudes toward technology in the top tier of a three-tier structure of attitudes (also including attitudes toward the product and the company), relating to online buying behavior. Indeed, Modahl (2000) identifies consumers' attitudes toward technology as the single most important determinant of personal computer buying and Internet shopping. Furthermore, it has been suggested that a low level of *technology anxiety* could be a better predictor of using self-service technologies, such as the Internet, than demographic variables (Meuter et al., 2003). The level of technology anxiety was also found to influence the overall level of satisfaction, the intention to use the technology again, and the likelihood of engaging in positive word of mouth (Meuter et al., 2003). Social influence is found to be another important determinant of Internet users' intentions to purchase online, second only to the attitude towards Web purchasing (Athiyaman, 2002).

Attitudes toward technology also influence consumers' perceptions of the risks associated with online purchasing. As Lee and Tan (2003) state, "the perceived product and service failure rates will be higher under online shopping than under in-store shopping" (p. 879).

Risk Perceptions

Risk perceptions are considered by many researchers as a strong barrier to online shopping (Kolsaker & Payne, 2002). Specifically, four types of perceived risk may prevent browsers from becoming shoppers: (1) financial risk; (2) time/convenience risk; (3) product performance risk; and (4) privacy concerns (compare Forsythe & Shi, 2003). Other risks associated with the Internet include credit card fraud, inability to touch and feel something before buying it, and problems of returning goods that fail to meet expectations (Kau et al., 2003). Additionally, difficulties in building trust in an online site may affect customers' willingness to purchase and to return to the site (Lynch, Kent & Srinivasan, 2001). A money back guarantee would therefore be the most important risk reliever, followed by offering a well-known brand and a price reduction (Van den Poel & Leunis, 1999).

Female consumers are reported to perceive online shopping as more risky than their male counterparts (Brown et al., 2003; Goldsmith & Bridges, 2000). Importantly, risk perceptions are also related to the context of the product being purchased; for instance, perceptions of security risk are highest for financial services (Montoya-Weiss, Voss & Grewal, 2003). Indeed, product type is an important influence on other consumer-related factors, such as the need to handle the product prior to purchase.

Need to Handle Product

While products' sensory attributes may have a direct influence on the suitability of the Internet as a retail channel, many researchers have examined the need for a tactile input as a characteristic of the individual consumer. Since the early developments of e-commerce, researchers have identified a negative relationship of online shopping with how important consumers feel the need to feel and handle goods before purchase (Falk, Talarzyk & Widing, 1994). Of particular interest to practitioners is the notion that consumers' need to handle products prior to purchase can discriminate between adopters and nonadopters of Web retailing, in terms of their likelihood to purchase products online. Specifically, consumers' desire to handle merchandise is a critical issue in nonstore retailing adoption, and for many product types, those desiring the ability to handle a product prior to purchase have more negative attitudes toward online shopping (Fenech & O'Cass, 2001). Additionally, the need for tactile input has a negative impact on buying online, especially for women who, compared with men, appear to have a higher need for tactile input when making product evaluations (Citrin et al., 2003). Retailers that operate both as brick-and-mortar and clicks-and-mortar should therefore offer to the consumer the ability to have direct multisensory experiences that may be desirable for evaluating alternatives. Concepts related to the need to handle the product are consumers' orientation toward shopping and the value they expect to gain from such activity.

Shopping Orientation

In addition to the attitudes and perceptions already discussed, consumers' attitudes toward a nonstore environment, in general, and e-commerce, in particular, is found to be a significant influence on the usage of online retailers (Helander & Khalid, 2000). Specifically, the utility and value that consumers expect to derive from the online shopping experience are related to their likelihood of using the Web for their purchases (Lee & Tan, 2003).

Indeed, attitudes toward shopping and buying online are found to discriminate between online shoppers and nonshoppers (Goldsmith & Bridges, 2000). The former highly rate the convenience, selection, and time saving benefits of shopping online, whereas the latter are worried about security, loss of privacy, and not receiving the goods they have ordered.

According to Eastlick and Lotz (1999), shopping orientation could discriminate between users and nonusers of nonstore environments. The consumer behavior literature typically discusses shopping orientation in terms of *economic* versus *recreational* orientation (Bellenger & Korgaonkar, 1980) or *hedonic* versus *utilitarian* (Babin, Darden & Griffin, 1994). If convenience is the main benefit expected from shopping online, then the Internet may best suit utilitarian consumers who perceive the act of shopping as a necessary task to be performed as fast as possible. On the other hand, consumers who enjoy the act of shopping, per se, and thus exhibit a hedonic attitude, could also derive an added value from the Internet, even though of a different nature. Thus, a hedonic consumer could enjoy buying online because of the novelty of this form of shopping, the fun, and the curiosity of a new way of purchasing (Dall'Olmo Riley, Scarpi & Manaresi, 2003).

Indeed, Fenech and O'Cass (2001) found that recreational shoppers, who enjoy shopping as a leisure activity, have a more positive attitude toward Web retailing than economic shoppers, who either dislike the experience of shopping or feel neutral toward it. Sénécal, Gharbi, and Nantel (2002) indicate a positive relationship between the experience of flow and the hedonic value of consumers' online shopping experiences. However, the two orientations may coexist: the two largest clusters of Internet shoppers were found to be the recreational and the price oriented consumers (Brown et al., 2003; Scarpi, 2004).

A possible implication for managers of brick-and-click stores is to raise the hedonic potential of their online shop, while also promoting the convenience of in-home shopping.

Retail Factors: Implications for Managers

As outlined in Figure 1 and as discussed above, product-related factors interact with consumer-related factors in determining the online purchase process. Product and consumer factors act both as constraints and as determinants of e-marketing strategies and tactics. At the same time, retailer factors can play an important role in easing the difficulties arising from product characteristics, reducing risk perceptions, and providing a shopping environment which not only facilitates purchase but also delivers enjoyment and satisfaction. In this perspective throughout the next sections, we discuss the main implications and provide some actionable suggestions for e-marketing managers. First, we deal with the *strategic concern* of the relationship between off-line and online provision, then we address the *tactical issues* of (1) Web site content and layout; (2) technology; (3) the provision of risk relievers; and (4) the provision of incentives.

Strategic Concern: Relationship between Off-Line and Online Retail Provision

There is strong evidence from both commercial (BRIE-IGCC E-Economy Project, 2001; Key Note, 2003) and academic research (Burke, 2002) that consumers are more willing to shop from online retailers that are multichannel, rather than Internet only, especially if they have already dealt with them through other channels. Some shoppers may buy online but still want to pick up their purchase at the closest store or may want to return merchandise purchased online to a store rather than by post (Burke, 2002).

There are advantages to be gained in pursuing a multiple distribution channel strategy (Citrin et al., 2003). Apart from the obvious branding implications, traditional brick-and-mortar stores fulfill an important function in allowing prospective Web shoppers with the opportunity to evaluate products directly before online purchase. Even for services, a face-to-face interaction with the service provider may facilitate an ensuing Internet purchase, again by providing an important tangible cue.

The challenge for retailers is to integrate the channels to allow flexibility, even though the process of managing multiple distribution systems may be difficult and risky. For instance, a failed attempt to purchase from a retailer online may have a negative effect on purchases from the same retailer off-line (Boston Consulting Group, 2001). Similar issues of multichannel integration apply to mail order companies offering their products both by catalog and online.

The general consensus from the literature is that traditional retailers should not abandon their brick-and-mortar operations in favor of purely Web-based sites (Burke, 2002; Citrin et al., 2003). On the other hand, pure Internet retailers may want to offer to their customers an alternative to placing the order online. For instance, *Dell.com* offered the opportunity to place an order by phoning a salesman in the call center. This strategy prevents the loss of customers who are worried about credit card security or uncertain of what to buy (Chaffey, 2000).

Tactical Factors

Web Site Content and Layout

The first and probably most quoted tactical consideration concerns the content and layout of the Web site. The basic rationale is that if the electronic shopping environment evokes positive affect, consumers will perceive greater value from their time, which in turn will serve as a reward, encouraging further patronage. Indeed, inadequate navigability of the site is a commonly quoted mistake, alongside with insufficient variety of products and brand selection (c.f., Cowles et al., 2002).

Good site design includes having fast, uncluttered, and easy to navigate sites (Szymanski & Hise, 2000). Consumers attempting to shop online should find the Web site easy to use (Burke, 1997; Chaffey, 2000; Goldsmith & Bridges, 2000) and simple in content, context and infrastructure (Wikström et al., 2002).   A simple layout, with easy sequential steps and straightforward information about 'hidden charges' (e.g., delivery cost or tax) will not only prevent the 'abandoned cart' syndrome but will also increase consumers' perceptions that the Web is a useful and convenient way of shopping, hence attracting new users.

The online shopping interface should conform to the different requirements of consumer-buying situations. For instance, elements from the physical world (e.g., a 'shopping basket') should be combined with 'virtual' elements (e.g., making it easier to find information) (Wikström et al., 2000). Detailed product information and expert evaluations should be offered to online shoppers of infrequently purchased products such as electronic goods, whilst a fast shopping experience with instant fulfillment should be offered to grocery buyers.

In a routine buying purchase situation, the site should 'remember' anything that requires repetition, such as passwords, credit card numbers, addresses, shopping lists, and size information (c.f., Moon, 2000). Besides making online shopping faster and easier, 'remembering' what customers want (e.g., a grocery shopping list) can also 'tie them in' and prevent them from switching to a competitor's site.

At the same time, a 'pleasurable' online shopping experience should satisfy the requirements of the consumers who have a more hedonistic orientation to shopping, stimulating exploration, curiosity and entertainment. Similarly to the role of atmospheric in offline retailing, the use of colors, music and other sensory features of the Web site should be carefully studied and selected, not to interfere with, but to enhance the shopping experience. Users' gratification from using the Web site is in fact suggested as an important driver of repeated use of the site (Joines et al., 2003; Szymansky & Hise, 2000). For instance, the personalized "hedonic" site management of *landsend.com* has attracted about 15 million visitors and doubled Internet sales from $61 million in just one year, attracting a 20% share of new customers (Cross & Neal, 2000).

Brick-and-mortar retailers should try to re-create online the same atmosphere experienced by consumers when they shop offline (Citrin et al., 2003). Congruity between the offline and online shopping experience is extremely important not only in 'hedonic' terms, but also to ensure consistency in the branding effort across channels (Earl, 2000). This may be particularly important for luxury brands, as shown by the efforts in this respect undertaken by companies such as *tiffany.com* and *giorgioarmani.com*.

Furthermore, satisfaction online is more positive when consumers perceive online stores to offer superior product assortments, satisfying their need for variety and stimulating their curiosity-exploration (Szymanski & Hise, 2000). Superior assortments therefore have a positive effect both on "hedonic" and "utilitarian" value. This becomes especially true if consumers desire goods not widely distributed, as the space available at brick-and-mortar stores is limited, but the space online is virtually infinite. For instance, whilst a large offline book superstore carries about 150,000 titles (Bianco, 1997), *Amazon.com* carries millions of titles.

While the Internet represents a tremendous opportunity, retailers new to e-commerce must consider the challenge of the technology-based differences between the two channels.

Technology

Technological issues are strictly related not only to the Web site's content and layout but also to the effectiveness in attracting and retaining customers. Managers should avoid the use of nonessential technological complexity in order to minimize the technology anxiety of some users (Meuter et al., 2003). A possible tactic would be to create a Web site with different levels of complexity to cater for different competency levels of browsers. Those unfamiliar with the Web might be offered a fast track easy step option for the purchase of a standardized offering, whereas expert shoppers may be given the opportunity to customize their purchases via a more complex route to the check out (compare Moon, 2000). For instance, *Nike.com* customers are able to customize the model, style, and color of the shoes they buy online but only if they wish to do so.

Joines, Scherer, and Scheufele (2003) advocate that Web pages should not follow a typical paper catalog format, but the interactive features of the Web should be exploited fully. For instance, all Web sites should provide consumers with the opportunity of getting in touch with a representative of the company, and the answer to any questions should come speedily either by e-mail or by phone. A tracking system similar to the one provided by companies success-fully operating online, like *Vertbaudet.co.uk* and *Amazon.com*, should be used to keep the purchaser up-to-date (by e-mail) with the status of the order, especially if an item is out of stock or there are any delays in the delivery. This would also help to overcome possible risk perceptions. Indeed, logistics problems and delivery delays or failures are often mentioned as a serious disincentive to Internet shopping, especially for first time buyers (Cowles et al., 2002; Scarpi, 2004).

The virtual features of the Internet should play a fundamental role in providing customers with tangible cues that can overcome the need of touching and feeling the product or of interacting face-to-face with a service provider. Virtual experience of the good can be granted with technology, such as the virtual model feature available at *landsend.com*. For infrequently purchased goods or new products, video streaming could be used to show the product in operation, illustrating its functions, characteristics, and use. For services, video streams can provide a virtual experience of a face-to-face encounter and illustrate the process of delivery. Explanation of the feature of the service product (e.g., a holiday destination or home content insurance) can also be illustrated more effectively by a combination of video images and text, rather than by text or still pictures alone. Additionally, video streams and music can also enhance the perceived hedonic value of the site.

Finally, the interactivity and intelligence features of the Internet should be exploited fully to customize products and services according to the requirements of individual consumers. Besides cookies, Web logs should be used to analyze how customers came to the site, the content they accessed and for how long, and how they respond to special offers. For instance, the UK e-retailer *Jungle.com* uses this technique to change the offers of its home page if customers are not responding (Chaffey, 2000). Technology should also be used to anticipate what customers want. For example, *Amazon.com* and other e-retailers use collaborative filtering systems (e.g., Firefly) which compare an individual customer's preferences with those of the entire user base. This enables Amazon to recommend new selections to each customer, based on the products bought by other users who have similar preferences (Moon, 2000). E-grocers could also use this technique to cross-sell products and to provide tailor-made special offers to individual customers, based on their usual shopping basket.

Risk Relievers

As discussed above, risk perceptions are a strong deterrent to Internet shopping. Managers should consider all or a combination of the following risk relievers.

Brand reputation is advocated by many researchers as the most important risk reliever (Citrin et al., 2003; Lee & Tan, 2003; Van den Poel & Leunis, 1999), particularly in product performance, financial, and psychological terms. Established off-line and catalog retailers will have an obvious advantage over newcomers, and they should exploit their reputation fully in their advertising. On the other hand, newcomers to electronic retailing, who lack an established reputation, can reduce consumers' perceived risk by carrying well-known brands (Lee & Tan, 2003). One alternative for start-ups in e-commerce would be to establish some kind of partnership with an existing traditional retailer. For instance, in the UK, the newcomer to e-grocery business *Ocado.com* operates in partnership with the well-known supermarket chain Waitrose. Money back guarantees, trust marks, free trials, clear explanations of the return policies, toll-free complaint hotlines, and possibly price reductions should also be offered, alongside third-party reviews and endorsements (Burke, 1997).

Incentives

As already mentioned, experience with Web use and with online purchasing is an important influence on the future use of the Web as a shopping medium. Hence, managers should consider providing incentives to attract first time buyers and to encourage customers' loyalty to the site. Vouchers, discounts, free gifts, and special offers (e.g., free delivery and returns) are possible incentives for both first time buyers and existing customers, especially if they have not purchased for a while. For instance, the UK e-grocer *Ocado.com* offers money off to regular customers as an incentive to shop and book a delivery slot on certain days and times.

Finally, e-retailers should use appropriate service recovery actions to attract dissatisfied customers back to the site. For instance, The Great Little Trading Company recently e-mailed all their online customers apologizing and explaining the reasons for delays and mistakes in deliveries which had occurred the previous season, while reassuring them of the steps they had undertaken to rectify the problem.

Conclusions

This chapter has provided suggestions of many possible ways of reducing or even overcoming the barriers preventing consumers from increasing the proportion of purchases they conduct online. Conversely, firms may decide to focus more on providing information and better services over the Internet, rather than attempting to sell. For instance, online insurance sales make up only about 1% of the $240 billion insurance market in the U.S. with insurers investing in providing additional services and information, rather than selling online (Bradford, 2001).

Clarity of objectives (e.g., selling or providing information and the balance between online and off-line provision) are paramount to any successful online strategy.

Acknowledgment

The authors would like to thank the anonymous reviewers for constructive and helpful comments.

References

Alba, J., Lynch, J., Weitz, B., Janiszewski, C., Lutz, R., Sawyer, A., & Wood, S. (1997, July). Interactive home shopping: Consumer, retailer, and manufacturer incentives to participate in electronic marketplaces. *Journal of Marketing, 61*, 38-53.

Athiyaman, A. (2002). Internet users' intention to purchase air travel online: An empirical investigation. *Marketing Intelligence and Planning, 20*(4), 234-242.

Babin, B.J., Darden, W.R., & Griffin, M. (1994). Work and/or fun: Measuring hedonic and utilitarian shopping value. *Journal of Consumer Research, 20*, 644-656.

Balabanis, G., & Reynolds, N.L. (2001). Consumer attitudes towards multi-channel retailers' Web sites: The role of involvement, brand attitude, Internet knowledge and visit duration. *Journal of Business Strategies, 18*(2), 105-115.

Bellenger, D.N., & Korgaonkar, P.K. (1980). Profiling the recreational shopper. *Journal of Retailing, 56*, 77-92.

Bellman, S., Lohse, G.L., & Johnson, E.J. (1999). Predictors of online buying behavior. *Communications of the ACM, 42*(12), 32-38.

Bianco, A. (1997). Virtual bookstores start to get real. *Business Week, 27*, 148-149.

Boston Consulting Group. (2001, July 17). An Integrated Multichannel Approach Creates Competitive Advantage for European Retailers, *New report from the Boston Consulting Group reveals the internet profoundly influences European consumers' offline shopping behavior.* http://www.bcg.com/News_Media/news_media_releases.jsp?id=943

Bowen, J. (1990). Development of a taxonomy of services to gain strategic marketing insights. *Journal of the Academy of Marketing Science, 18*, 43-49.

Bradford, M. (2001). Internet seen as service provider, not sales tool. *Business Insurance,35*(8), 22-23.

Brie-IGCC E-Economy Project. (2001). Tracking a transformation. *E-commerce and the terms of competition in industries.* Washington, DC: Brookings Institution Press.

Brown, M., Pope, N., & Voges, K. (2003). Buying or browsing? An exploration of shopping orientations and online purchase intention. *European Journal of Marketing, 37*, 1666-1684.

Burke, R.R. (1997). Do you see what I see? The future of virtual shopping. *Journal of the Academy of Marketing Science, 25*, 352-360.

Burke, R.R. (2002). Technology and the customer interface: What consumers want in the physical and virtual store. *Journal of the Academy of Marketing Science, 30*, 411-432.

Chaffey, D. (2000). Achieving Internet marketing success. *The Marketing Review, 1*, 35-59.

Citrin, A.V., Stern, D.E., Spangenberg, E.R., & Clark, M.J. (2003). Consumer need for tactile input. An internet retailing challenge. *Journal of Business Research, 56*, 915-922.

Cowles, D.L., Kiecker, P., & Little, M.W. (2002). Using key informant insights as a foundation for e-retailing theory development. *Journal of Business Research, 55*, 629-636.

Cross, R., & Neal, M. (2000). Internet retailers are creating their own brand of service. *Direct Marketing, 63*(2), 20-23.

Dall'Olmo Riley, F., Scarpi, D., Manaresi A., Lago, U., Pezzini, F. (2003, May 21-23). Hedonism on the wire? A Cross Country Comparison. *Proceedings of the 32nd European Marketing Conference*. Glasgow, UK. *http://www.marketing.strath.ac.uk/emac2003/proceedings/authors-s.htm*

Davis, F.D., & Venkatesh, V. (1996). A critical assessment of potential measurement biases in the technology acceptance model: Three experiments. *International Journal of Human-Computer Studies, 45*, 19-45.

Degeratu, A., Rangaswamy, A., & Wu, J. (2000). Consumer choice behavior in online and traditional supermarkets: The effects of brand name, price and other search attributes. *International Journal of Research in Marketing, 17*(1), 55-78.

Dieringer Research Group. (2003). American interactive consumer survey. In L. Mazur (Ed.), *Brands are not doing justice to rise of the Web*, Marketing, 23 October, 2003.

Earl, M.J. (2000). Evolving the e-business. *Business Strategy Review, 11*, 33-38.

Eastlick, M.A., & Lotz, S. (1999). Profiling potential adopters and non-adopters of an interactive electronic shopping medium. *International Journal of Retail & Distribution Management, 27*(6), 209-223.

Falk, H., Talarzyk, W., & Widing II, R.E. (1994). Retailing and online consumer information services. *International Journal of Retail and Distribution Management, 22*, 18-23.

Fenech, T. (2000). Attitude and security do count for shopping on the World Wide Web. *Proceedings of ANZMAC 2000 Conference: Visionary Marketing for the 21st Century: Facing the Challenge*. Australia.

Fenech, T., & O'Cass, A. (2001). Internet users' adoption of Web retailing: User and product dimensions. *Journal of Product & Brand Management, 10*, 361-381.

Forsythe, S.M., & Shi, B. (2003). Consumer patronage and risk perceptions in Internet shopping. *Journal of Business Research, 56*, 867-875.

Girard, T., Silverblatt R. & Korgaonkar, P. (2002). Influence of product class on preference for shopping on the Internet. *Journal of Computer Mediated Communication, 8*(1). Retrieved October 8, 2004, from *http://www.ascusc.org/jcmc/Vol8/issue1/girard.html*

Goldsmith, R.E,. & Bridges, E. (2000). E-tailing vs. retailing. Using attitudes to predict online buying behavior. *Quarterly Journal of Electronic Commerce, 1*, 245-253.

Goldsmith, R.E., & Goldsmith, E.B. (2002). Buying apparel over the Internet. *Journal of Product and Brand Management, 11*(2), 89-102.

Hall, M. (2003, September 21). Online living: Still a few clicks away from becoming a reality. *Sunday Times, http://business.timesonline.co.uk/ article/0,,8209-1371018,00.html*

Hart, C., Doherty, N., & Ellis-Chadwick, F. (2000). Retailer adoption of the internet – Implications for retail marketing. *European Journal of Marketing, 34*, 954-974.

Harvin, R. (2000). In Internet branding, the offline have it. *Brandweek, 41*, January 24, 30-31

Helander, M.G., & Khalid, H.M. (2000). Modeling the customer in electronic commerce. *Applied Ergonomics, 31*, 609-619.

Joines, J., Scherer, C., & Scheufele, D. (2003). Exploring motivations for consumer Web use and their implications for e-commerce. *Journal of Consumer Marketing, 20*(2), 90-108.

Kau, A.K., Tang, Y.E., & Ghose, S. (2003). Typology of online shoppers. *Journal of Consumer Marketing, 20*(2), 139-156.

Key Note. (2003). *Home shopping.* Hampton: Key Note.

Klein, L.R. (1998). Evaluating the potential of interactive media through a new lens: Search versus experience goods. *Journal of Business Research, 41*, 195-203.

Kolsaker, A., & Payne, C. (2002). Engendering trust in e-commerce: A study of gender-based concerns. *Marketing Intelligence and Planning, 20*(4), 206-214.

Lee, M.Y., & Johnson, K.K.P. (2002). Exploring differences between Internet apparel purchasers, browsers and non-purchasers. *Journal of Fashion Marketing and Management, 6*(2), 146-157.

Lee, K.S., & Tan, S.J. (2003). E-retailing versus physical retailing: A theoretical model and empirical test of consumer choice. *Journal of Business Research, 56*, 877-886.

Lovelock, C.H. (1984). *Services Marketing.* Englewood Cliffs, NJ: Prentice Hall.

Lynch, P.D., Kent, R.J., & Srinivasan, S.S. (2001, May-June). The global Internet shopper: Evidence from shopping tasks in 12 countries. *Journal of Advertising Research*, 15-23.

Mazur, L. (2003, October 2). Brands are not doing justice to rise of the Web. *Marketing*.

Meuter, M.L., Ostrom, A.L., Bitner, M.J., & Roundtree, R. (2003). The influence of technology anxiety on consumer use and experiences with self-service technologies. *Journal of Business Research*, *56*, 899-906.

Modhal, M. (2000). *Now or never: How companies must change today to win the battle for Internet consumers*. New York: Harper Collins.

Montoya-Weiss, M.M., Voss, G.B., & Grewal, D. (2003). Determinants of online channel use and overall satisfaction with a relational, multichannel service provider. *Journal of the Academy of Marketing Science, 31*, 448-458.

Moon, Y. (2000). Interactive technologies and relationship marketing strategies. *Harvard Business School Note, 9-599-101*.

Nelson, P.J. (1974). Advertising as information. *Journal of Political Economy, 82*, 729-754

Peterson, R.A., Balasubramanian, S., & Bronnenberg, B.J. (1997). Exploring the implications of the Internet for consumer marketing. *Journal of the Academy of Marketing Science, 25*, 329-346.

Scarpi, D. (2004, May). A comparison of consumers' orientation towards electronic shopping. Proceedings of the 33rd European Marketing Conference, EMAC, Spain.

Sénécal, S., Gharbi, J.-E., & Nantel, J. (2002). The influence of flow on hedonic and utilitarian shopping values. *Advances in Consumer Research, 29*, 483-484.

Szymanski, D.M., & Hise, R.T. (2000). E-satisfaction: An initial examination. *Journal of Retailing, 76*, 309-322.

Van den Poel, D., & Leunis, J. (1999). Consumer acceptance of the Internet as a channel of distribution. *Journal of Business Research, 45*, 249-256.

Wikström, S., Carlell, C., & Frostling-Henningsson, M. (2002). From real world to mirror world representation. *Journal of Business Research, 55*, 647-654.

Chapter IV

eCRM:
Understanding Internet Confidence and the Implications for Customer Relationship Management

Terry Daugherty, University of Texas at Austin, USA

Matthew Eastin, Ohio State University, USA

Harsha Gangadharbatla, University of Texas at Austin, USA

Abstract

As we enter the 21st century many firms implementing Customer Relationship Management strategies have turned to the Internet as a primary means for collecting consumer data. Consequently, understanding when consumers are willing to comply with data requests has become increasingly important to e-marketers. However, current research has failed to explore how self-confidence with using the Internet impacts a consumer's willingness to provide personal information online. Therefore, this chapter reports findings from an online consumer panel survey designed to investigate how perceived Internet confidence influences consumer attitudes toward divulging personal information and their willingness to comply with data requests online (n=500). The results largely support the notion that enhancing Internet confidence can lead to more favorable attitudes toward information requests and increased willingness to provide information.

Introduction

With over half of the U.S. population online and a growth rate of in excess of two million users per month, the Internet has become an important mainstream medium (NTIA, 2002). This widespread adoption signifies the convergence of two long-term trends in business: the rapid expansion of the information economy and the rise of customer service over the Web (Rust & Kannan, 2002). Accordingly, there has been radical change in the field of marketing as many companies recognize the potential of this unique medium for efficiently delivering targeted messages, generating sales, and facilitating two-way communication with consumers. One such change has been the emergence of the Internet as a powerful electronic customer relationship management (eCRM) tool.

As we enter the 21st century, technological innovations have enabled marketers to collect large amounts of information and build databases full of consumer profiles (Pardun & Lamb, 1999). In fact, many firms implementing customer relationship management (CRM) today have turned to the Internet, in particular the Web, as a primary means for collecting customer information (Masci, 1999). When you consider as many as 90 million Americans use the Internet daily, it is not surprising to learn that 97% of all commercial Web sites attempt, in some form or another, to collect personal information from their visitors (Federal Trade Commission, 2000). The type of information Web sites collect in order to build their databases for consumer research and relationship marketing varies from gender, age, martial status, educational level, ethnicity, occupation, household income, Internet usage, and even buying patterns (Turban, Lee, King & Chung, 2000). Academicians and practitioners alike have been vociferous in pointing out the importance of such data gathering with regard to the successful implementation of any CRM strategy (Abbott, Stone

& Buttle, 2001; Hall, 2001; Zhang, Wang & Chen, 2000). The reason the Internet has become so important is because eCRM data acquisition often streamlines record entry and arrangement used during analyses to profile existing, as well as prospective, customers. Furthermore, eCRM enables a company to create the perception of one-to-one communication through interactivity with very little incremental cost, while maintaining the added advantage of mass marketing because of widespread adoption (Noyes, 2000).

With corporate implementation and adoption of eCRM ongoing, understanding when consumers are willing to comply with data requests and their subsequent thought processes while providing personal information online becomes increasingly important. One acknowledged explanation as to why consumers agree to provide marketers with personal information is when the obtained benefits (i.e., incentives or returns) exceed their anticipated risk (Gangadharbatla, Li & Edwards, 2002; Zhang et al., 2000). However, recent research has also indicated that an Internet user's personal beliefs and subsequent confidence with technology can also serve an influential role in understanding online behavior (LaRose & Eastin, 2002).

The individual beliefs formulated about one's ability to comprehend concepts and effectively use this medium reassures decision-making while also reinforcing actions online to serve as the basis for establishing a consumer's personal level of Internet confidence. Because customer expectations are shaped by technology in this information economy, the cognitive process of developing Internet confidence is a crucial component for understanding online behavior (Eastin & LaRose, 2000). Nevertheless, current research has failed to explore the impact Internet confidence has on a consumer's willingness to comply with data requests. This is surprising considering that e-marketing efforts rely so heavily on gathering personal information online. Therefore, the purpose of this chapter is to expand our theoretical understanding of e-commerce by introducing Internet confidence within the realm of CRM. Obviously, there are other factors discussed throughout this book that also impact a consumer's willingness to supply information online (e.g., privacy, trust, security, etc.), yet our exploratory focus remains to examine the influence of Internet confidence on consumer behavior. In the proceeding sections, we propose a hypothesis model for understanding Internet confidence, present research results testing this model, and conclude with eCRM implications.

Literature Review

Understanding eCRM

Customer relationship management is a long-term business philosophy focused on collecting, understanding, and utilizing cumulative customer information intelligently in order to continuously assess consumer needs (Varki, 2002). Not surprising, one of the foremost factors influencing the success or failure of any CRM strategy is the quality of the data collection and analysis (Abbott, et al., 2001; Farah & Higby, 2001; Masci, 1999; Zhang, et al., 2000). This is precisely why many firms implementing CRM today have turned to the Internet as a primary means for collecting customer information. Of course, this trend has led to a multitude of concerns with regard to how companies collect, buy, and sell one's personal information. As a result, over 2,000 Web sites now display some form of recognized privacy seal (i.e., TRUSTe, BBB Online, etc.) designed to alleviate apprehension over gathered personal information (Erlanger, 2004). Nevertheless, 52% of U.S. consumers remain very uncomfortable with providing personal information via the Web (Online Selling and eCRM, 2003).

The realization of the Internet as a CRM tool means the collection, dissemination, and commercial use of personal information is now easier to capture than ever before (Cavoukian & Tapscott, 1996). Accordingly, many firms today are turning to the Internet not just to establish an online identity but to also expand their ability to acquire customer information through online data collection (Masci, 1999). With corporate implementation and adoption of eCRM ongoing, understanding the factors that influence a consumer's willingness to provide information online becomes increasingly important. However, most attempts to gather information online are often rejected by Internet users because consumers are very concerned about providing personal information online (Camp, 1999; Caudill & Murphy, 2000). These concerns can stem from numerous factors, such as gender, age, occupation, and cultural background (Zhang et al., 2000, 2001), as well as one's personal understanding of technology (LaRose & Eastin, 2002). While trust, privacy, and security related issues remain top barriers to online data collection, the scope of this chapter is focused on understanding the effect of one's ability to use technology, referred to as Internet confidence.

Internet Confidence

While there have been numerous studies investigating the impact of self-confidence on consumer decision-making and behavior (Bearden, Hardesty & Rose, 2001; Bettman, Johnson & Payne, 1991; Fleming & Courtney, 1984), very little is known about the nature of self-confidence as it applies to e-commerce and especially one's willingness to provide personal information online. In fact, the construct of confidence can have different meanings depending on the context. For instance, it may refer to a person's trust in another, another person's ability to perform a task, or a person's judgment about a future event (Barbalet, 1998). In addition, the concept may also refer to one's belief in self-ability, which is the conceptualization used in this framework. The feeling of confidence in one's ability has been characterized as essential for any behavior to take place because this belief serves as a form of self-assurance (Dequech, 2000). With regard to using the Internet, personal confidence in one's ability to successfully understand, navigate, and evaluate content should alleviate doubts when providing information online corresponding with heightened beliefs about data collection practices. These beliefs formed reflect a consumer's perceived capability in using the Internet to accomplish tasks (Eastin & LaRose, 2000). Subsequently, as Internet confidence (i.e., beliefs) increases, then attitudes toward the object of those beliefs will also increase (Ajzen & Sexton, 1999, p. 118).

H1: Internet confidence will be positively associated with a consumer's attitude toward providing personal information online.

Earlier work conducted to understand technology behavior has applied Fishbein and Ajzen's (1975) Theory of Reasoned Action as an explanatory model. This framework considers the beliefs that an individual has about a behavior and the actions taken from those beliefs. While this approach has had success predicting the uses of computing technologies (Davis, Bagozzi & Warsaw, 1989), later studies indicate that modifications to the model were needed. For example, Ajzen's (1991) extension as the Theory of Planned Behavior suggests that an individual's perceived ability, or confidence, to perform a behavior could also play an important role in the adoption process, ultimately suggesting that the level of perceived complexity experienced from a behavior could influence the formation of internal perceptions (Ajzen, 1991).

More recently, Bandura's (1986, 2001) Social Cognitive Theory has been posited as a means for understanding the use of information technology (Compeau & Higgins, 1995; Compeau, Higgins & Huff, 1999; Eastin & LaRose, 2000; LaRose & Eastin, 2002). Social Cognitive Theory includes a complex triadic causal structure that establishes the development of competency and regulation of action (Bandura, 1986, 2001). Through the development of knowledge structures, cognitive models of action are created which guide behavior. Consequently, this cognitive guidance is a crucial component to the developmental stages of a behavior. One cognitive factor that has been characterized as an influential variable is perceived confidence, more commonly referred to as self-efficacy (Bandura, 1997).

As a primary self-regulatory mechanism, self-efficacy simply refers to the level of confidence an individual has toward a given behavior. As a central component of Social Cognitive Theory, this confidence represents the internal belief "in one's capability to organize and execute the courses of action required to produce given attainments" (Bandura, 1997, p. 3). To date, researchers have positively linked Internet confidence to online performance, prior experience, and Internet use (Eastin & LaRose, 2000). Furthermore, Eastin (2002) found that Internet confidence was positively related to e-commerce activities, such as online banking. Thus, it is reasonable to assume that confidence in one's ability to use the Internet will be positively related to their willingness to provide personal information online because internal beliefs are associated with actual behavior.

H2: Internet confidence will be positively associated with a consumer's willingness to provide personal information online.

Bearden and colleagues (2001) characterize consumer confidence as the extent to which individuals feel capable and secure with their decisions and behaviors. As one's confidence increases, a person is better equipped at making decisions and feels confident about the resulting behavior. However, behavior is more strongly associated with one's attitude toward that action when deliberate cognitive processing is involved (Fazio & Towles-Schwen, 1999, p. 99). Because Internet confidence does not represent a form of deliberate cognitive processing directly associated with the behavior in question (i.e., willingness to provide personal information online), attitude toward online data collection should exude a stronger direct influence on behavior than Internet confidence.

H3: Attitude toward online data collection will mediate the relationship between Internet confidence and a consumer's willingness to provide personal information.

Method

Sample

Five hundred adults were surveyed from an online panel for this study. The panel is an opt-in privacy protected subject pool recruited for Web-based research. For joining the panel and subsequently participating in research, panelists are eligible for both monthly and study specific cash prize drawings.

Design

A thirty-item questionnaire was developed and pretested on a small sample of academic professionals to insure clarity (additional data was collected yet not analyzed for this study). To assess each respondent's confidence using the Internet, a Likert-type scale derived from previous research was used (LaRose, Eastin & Gregg, 2001). Participants were asked to indicate their level of agreement with six statements using a 7-point scale (strongly disagree/strongly agree) designed to measure their confidence: towards understanding Internet terms or words, describing functions of the Internet, troubleshooting Internet problems, using the Internet to gather information, using the Internet for entertainment, and in their own Internet ability.

Respondents' attitudes toward providing information online was measured using six established 7-point items (unfavorable/favorable, negative/positive, annoying/pleasing, dislike very much/like very much, bad/good, disagreeable/agreeable) (Bruner, James & Hensel, 2001, p.84). Specifically, participants were asked to indicate their personal feelings about providing information online when approached by organizations seeking to collect consumer data. Finally, respondents indicated their willingness to comply with data requests using four established 7-point behavioral intention items (unlikely/likely, improbably/probably, uncertain/certain, impossible/possible) (MacKenzie & Spreng, 1992). In particular, participants were asked how likely they were to agree to an organization's request and provide personal information online.

Procedure

Selected participants were notified by e-mail that they had been identified for this study. The notification message explained the estimated time to complete the survey, offered a monetary incentive in the form of a cash drawing, and provided respondents with a URL to participate. By following the URL, participants were directed to a secure server and required to sign in using their ID and password specified when joining the panel.

Results

Data Analysis

The sample consisted of 306 men (61.2%) and 194 women (38.8%) with an average age of 40 (SD = 13.69). Furthermore, respondents were primarily Anglo (80.6%) and well educated with 54.8% holding a bachelor's degree, including 52.2% reporting household income levels of $75,000 or more. Ninety-four percent indicated they had been using the Internet for three or more years and, for the most part (56.6%), spent two hours or less a week online.

Reliability assessment was conducted on the Internet confidence, attitude, and behavior intention scales using Cronbach's Alpha (Internet confidence M=5.50, SD=1.36, α = .92; attitude M=3.83, SD=1.21, α = .94; behavior intent M=4.25, SD=1.22, a=.88) with all exceeding the generally accepted guideline of .70 (Hair, Anderson, Tatham & Black, 1998, p. 118). Composite measures for each of the scales were then constructed to represent the multiple items and used in the subsequent analysis to reduce measurement error.

Hypothesis Testing

Hypothesis 1 predicted that a participant's confidence in their ability to use the Internet is positively related to their attitude toward providing firms with personal information online. Using Pearson's correlation, a significant positive relationship was identified between participants' reported Internet confidence and their feelings about such data requests (r = .39, p < .001), supporting the hypothesis.

In turn, Hypothesis 2 predicted that a participant's confidence in their ability to use the Internet would also be positively related to their willingness to provide personal information online. The results indicate that a significant positive relationship was detected supporting the hypothesis as participants specified they were more likely to agree to a firm's data request as their Internet confidence increased ($r = .34$, $p < .001$).

Finally, Hypothesis 3 predicted that participant's reported attitude toward providing information online would mediate the influence Internet confidence exerts on a respondent's intent to comply with data requests. To verify this proposition, a mediation analysis was conducted as specified by Baron and Kenny (1986). In order for the mediation to occur, the following must hold: (1) Internet confidence must positively affect the mediator (attitude), (2) Internet confidence must positively affect the dependent variable (intent to comply by indicating their willingness to provide personal information), and (3) the mediator must positively affect the dependent variable when regressed in conjunction with the independent variable. Providing these conditions are met, the effect of the independent variable on the dependent variable must be less in the third step than in the second step (Baron & Kenny, 1986).

The first step indicates that Internet confidence positively influences participant's attitude toward providing information online ($\beta = .39$, $t(498) = 9.38$, $p < .001$, $R^2 = .15$). Furthermore, the second step demonstrates that Internet confidence positively influences intent ($\beta = .34$, $t(498) = 8.12$, $p < .001$, $R^2 = .12$). Finally, the third analysis supports attitude as a mediating variable between Internet confidence and consumer intent for supplying personal information online ($\beta = .65$, $t(2,497) = 18.64$, $p < .001$, $R^2 = .48$). Accordingly, the influence of Internet confidence on intent diminished when included in the analysis with attitude ($\beta = .09$, $t(2,497) = 2.53$, $p < .05$). Even though the strength of the relationship between Internet confidence and intent weakened considerably when regressed with attitude, there remained a significant association suggesting only partial mediation. While this residual relationship may be a product of the sample size, the finding is also supported within the framework of the Theory of Planned Behavior and Social Cognitive Theory. For instance, behavior is influence by both cognitive and personal factors with confidence impacting outcomes as well as internal mechanisms that are reciprocally determined (Compeau & Higgins, 1995). Thus, while a direct relationship is observed between confidence and attitude (i.e., internal), a behavioral outcome is not totally unexpected.

Conclusions

The confirmation of the relationship between Internet confidence, attitude, and behavior intent validates the hypothesized model and signifies the potential importance of these factors within eCRM strategies. Furthermore, the results largely support the notion that enhancing confidence toward using the Internet could potentially impact one's attitude toward online information requests and increase compliance for data requests. The mediation analysis confirms this established theoretical relationship between attitude and behavior. Specifically, respondents' attitude toward providing personal information online significantly predicts their behavior intention when information is requested. While a significant direct relationship was observed between Internet confidence and willingness to provide personal information when controlling for attitude, the observed effect was small and not completely unexpected.

By recognizing the contribution of Internet confidence toward eCRM implementation, marketers may be able to avoid data loss and improve strategic decision-making. For instance, if a firm's target market includes relatively inexperienced Internet users, then utilizing simple methods, such as one-step registration, could increase compliance for data collection requests. The premise is that simple instructions or requests are less likely to result in "drop outs" because consumer ability is never challenged. However, when the target market is considered an experienced Internet user, alternative strategies designed to enhance user control, such as the ability to customize a Web site or provide in-depth site navigation, will correspond with ability ultimately enhancing confidence. Furthermore, opt-in and name-removal site mechanisms may increase consumer control and awareness of the data collection process with regard to personal information. If these features or experiences are conclusive, a positive transference effect could possibly lead to increases in Internet confidence. Another strategy for increasing Internet confidence within complex Web sites is to provide an easy to follow tutorial or walkthrough. Because the degree of difficulty can vary greatly among Web sites, automated prompts or instructions that allow you to practice tasks or experience features before committing information may reduce anxiety associated with supplying sensitive information online. Interestingly, this act can also serve to build trust, which is then reinforced by third-party certifications and privacy seals. Similarly, a more proactive approach for matching consumer confidence before attempting to solicit data requests would be to segment consumers via a brief survey. While this would essentially require consumers' to comply with a data request, the instrument could be administered on the first site visit with the results stored in a cookie on the consumer's system. Depending on the responses, a firm could identify a consumer's level of Internet confidence and tailor specific site content, navigation, and information requests accordingly.

Regardless of the tactic used, in order to promote an optimal level of Internet confidence the goal should be to match the appropriate type of data collection request with the consumer's ability. Ultimately, the continued advancement of the Internet as an eCRM tool is important as it enhances a firm's ability to provide a customer-centric approach to their marketing efforts. Keep in mind though that online data collection may be affected by a variety of factors, and the eCRM industry as a whole should work toward building user confidence in the medium.

Research exploring the impact of Internet confidence has immediate theoretical and industry implications as marketers are caught in a technological information society continuing to advance. The findings presented in this chapter represent an initial attempt to understand the relationship between Internet confidence and a consumer's willingness to provide personal information online. With consumer adoption of the Internet and corporate implementation of eCRM ongoing, understanding the variables that influence a consumer's willingness to provide information becomes increasingly important.

Inherent within any research are potential limitations that affect the overall validity and reliability of the results. With regard to this chapter, a few limitations should be considered when interpreting the findings presented. The first concern involves error resulting from the sampling method. When a purposive sample is utilized, the technique obviously limits the external validity of the results. Nevertheless, the objective of this chapter was not to generalize findings to the entire Internet universe. Rather, the intent was to verify a relationship between the proposed theoretical constructs. A second concern involves the possibility of measurement error. Because a survey was used, several assumptions were made regarding the terminology of the questionnaire. Furthermore, attitudes toward providing information online and willingness to do so were ascertained in a very general context rather than inquiring about specific data collection techniques. While attitudinal scales may not reflect actual data entry behavior, the intent to provide information online was assumed to represent a suitable substitute. Therefore, future research should explore the influence of Internet confidence across multiple data collection types, markets and methods, such as e-mail, Web sites, unobtrusive practices, business-to-business segments, and controlled experiments. Another concern focuses on the level of Internet confidence respondents reported. With 85% of the scores falling above the midpoint of the Internet confidence scale, a ceiling effect is very likely with these results simply reflecting high levels of Internet confidence. This of course could be a by-product of using an online consumer panel compared to surveying Internet novices. However, the use of online panels is expected to account for 50% of all marketing research by 2005, signifying the mainstream nature of this resource (James, 2000). Regardless, the aim of this study was not to postulate outcome differences for varying levels of confidence but to first confirm whether a relationship between Internet confidence, attitude, and behavior intent exists. A final limitation to this study is the deliberate exclusion of issues, such as trust, privacy, and security, that play an important role in an individual's willingness to provide information online. While a significant relationship was detected for Internet confidence, this construct is undeniably a single influence among multiple variables. However, this isolated investigation also demonstrates the importance of considering Internet confidence as a contributing factor, which numerous other studies investigating eCRM have failed to explore.

Despite these limitations, the chapter is successful in advancing our initial understanding of Internet confidence with additional paths of research potentially leading to important findings in this new area. For instance, more theoretical research is needed designed to explore the unique and distinctive characteristics that distinguish between levels of Internet confidence. In addition, individual differences, such as personality characteristics, may not only enhance our understanding of how elevated levels of Internet confidence forms but also identify the rate at which belief in one's ability emerges. The bottom line is that we know very little about the intricacies of this construct, and continued research in this area is not only essential to our understanding of eCRM but also significant for e-commerce in general.

Acknowledgments

The authors would like to thank Wei-Na Lee and the Virtual Consumer Research Group in the Department of Advertising at The University of Texas at Austin for support of this research.

References

Abbott, J., Stone, M., & Buttle, F. (2001). Customer relationship management in practice: A qualitative study. *Journal of Database Marketing, 9*(1), 24-34.

Ajzen, I. (1991). The theory of planned behavior. *Organizational Behavior and Human Decision Process, 50*, 179-211.

Ajzen, I., & Sexton, J. (1999). Depth of processing, belief congruence, and attitude-behavior correspondence. In S. Chaiken & Y. Trope (eds.). *Dual-process theories in social psychology* (pp. 117-138). New York: Guilford Press.

Bandura, A. (1986). *Social foundations of thought and action: A social cognitive theory*. Englewood Cliffs, NJ: Prentice Hall.

Bandura, A. (1997). *Self-efficacy: The exercise of control*. New York: W.II. Freeman.

Bandura, A. (2001). Social cognitive theory of mass communication. *Media Psychology, 3*, 265-299.

Barbalet, J.M. (1998). *Emotions, social theory, and social structure: A macrosociological approach*. Cambridge: Cambridge University Press.

Baron, M., & Kenny, A. (1986). The moderator-mediator variable distinction in social psychological research: Conceptual, strategic, and statistical considerations. *Journal of Personality and Social Psychology, 51*(6), 1173-1182.

Bearden, W.O., Hardesty, D.M., & Rose, R.L. (2001, June). Consumer self-confidence: Refinements in conceptualization and measurement. *Journal of Consumer Research, 28*, 121-134.

Bettman, J.R., Johnson, E., & Payne, J.W. (1991). Consumer decision-making. In T.S. Robertson & H.H. Kasssarjian Harold (eds.), *Handbook of consumer behavior* (pp. 54-80). Englewood Cliffs, NJ: Prentice Hall.

Bruner, G.C., James, K.E., & Hensel, P.J. (2001). *Marketing scales handbook: A compilation of multi-item measures: Vol. 3*. Chicago: American Marketing Association.

Camp L.J. (1999). Web security and privacy: An American perspective. *The Information Society, 15*, 249-256.

Caudill E.M., & Murphy P.E. (2000). Consumer online privacy: Legal and ethical issues. *Journal of Public Policy and Marketing, 19*(1), 7-19.

Cavoukian, A., & Tapscott, D. (1996). *Who knows:Safeguarding your privacy in a networked world.* New York: McGraw-Hill.

Compeau, D., & Higgins, C. (1995). Computer self-efficacy: Development of a measure and initial test. *MIS Quarterly, 19*, 189-211.

Compeau, D., Higgins, C., & Huff, S. (1999). Social cognitive theory and individual reactions to computing technology: A longitudinal study. *MIS Quarterly, 23*, 145-158.

Davis, F.D., Bagozzi, R.P., & Warsaw, P.R. (1989). User acceptance of computer technology: A comparison of two theoretical models. *Management Science, 35*, 983-1003.

Dequech, D. (2000). Confidence and action: A comment on Barbalet. *Journal of Socio-Economics, 29*(6), 503-516.

Eastin, M.S. (2002). Diffusion of e-commerce: An analysis of the adoption of four e-commerce activities. *Telematics and Informatics, 19, 251-267.*

Eastin, M.S., & LaRose, R.L. (2000). Internet self-efficacy and the psychology of the digital divide. *Journal of Computer-Mediated Communication, 6.* Retrieved October 11, 2004, from *http://www.ascusc.org/jcmc/vol6/*

Erlanger, L. (2004). Should you trust TrustE? *PC Magazine, 23*(3), 59.

Farah, N.B., & Higby, M.A. (2001). E-commerce and privacy: Conflict and opportunity. *Journal of Education for Business, 76*(6), 303-307.

Fazio, R.H., & Towles-Schwen, T. (1999). The MODE model of attitude-behavior processes. In S. Chaiken & Y. Trope (Eds.)., *Dual-process theories in social psychology* (pp. 97-116). New York: Guilford Press.

Federal Trade Commission. (2000). Privacy online: Fair information practices in the electronic marketplace. *A Report to Congress.* Retrieved October 11, 2004, from *http://www.ftc.gov/reports/privacy3/index.htm*

Fishbein, M., & Ajzen, I. (1975). *Belief, attitude, intention, and behavior: An introduction to theory and research.* Reading, MA: Addison-Wesley.

Fleming, J., & Courtney, B.E. (1984, February). The dimensionality of self-esteem II: Hierarchical facet model for revised measurement scales. *Journal of Personality and Social Psychology, 46*, 404-421.

Gangadharbatla, H., Li, H., & Edwards, S. (2002). Incentive options for effective Web-based data collection: A conjoint analysis. In W.J. Kehoe & J.H. Lindgren (eds.), *Proceedings of 2003 American Academy of Advertising Conference, 38*, 145-154. Denver, CO: American Academy of Advertising.

Hair, J.F., Anderson, R.E., Tatham, R.L., & Black, W.C. (1998). *Multivariate data analysis* (5th ed.). Upper Saddle River, NJ: Prentice Hall.

Hall, O.P., Jr. (2001). Mining the store. *The Journal of Business Strategy, 22*(2), 24-27.

James, D. (2000, January). The future of online research. *Marketing News, 3*, 1-11.

LaRose, R., & Eastin, M.S. (2002). Is online buying out of control? Electronic commerce and consumer self-regulation. *Journal of Broadcasting & Electronic Media, 46*, 225-253.

LaRose, R., Eastin, M.S., & Gregg, J. (2001). Reformulating the Internet paradox: Social cognitive explanations of Internet use and depression. *Journal of Online Behavior, 1*(2). Retrieved October 11, 2004, from *http://www.behavior.net/JOB/*

MacKenzie, S.B., & Spreng, R.A. (1992, March). How does motivation moderate the impact of central and peripheral processing on brand attitudes and intentions? *Journal of Consumer Research, 18*, 519-529.

Masci, D. (1999). Internet privacy. In L.S. Sandra (ed.), *Issues for debate in American public policy* (pp. 175-191). Washington, DC: CQ Press.

Noyes, A. (2000, July). CRM in online retail: Four steps to success on the Web. *Unisys World Print*. Retrieved October 11, 2004, from *http://www.unisysworld.com/monthly/2000/07/4steps.shtml*

NTIA (National Telecommunications and Information Administration). (2002). A nation online: How Americans are expanding their use of the Internet. Retrieved October 11, 2004, from *http://www.ntia.doc.gov/ntiahome/dn/html/anationonline2.htm*

Pardun, C.F., & Lamb L. (1999). Corporate Web sites in traditional print advertisements. *Internet Research, 9*(2), 93-99.

Rust, R., & Kannan, P.K. (2002). Preface. In R. Rust & P.K. Kannan (eds.), *E-service: New directions in theory and practice* (pp. x-xii). New York: M.E. Sharpe.

Turban, E., Lee, J., King, D., & Chung, H.M. (2000). *Electronic commerce: A managerial rerspective*. New Jersey: Prentice Hall.

Varki, S. (2002). Real-time marketing in e-services. In R. Rust & P.K. Kannan (Eds.), *E-Service: New Directions in Theory and Practice* (pp. 154-167). New York: M.E. Sharpe.

Zhang Y., Wang, C.L., & Chen, J. Q. (2000). Consumer's responses to Web based data collection efforts and factors influencing the responses. *Journal of International Marketing and Marketing Research*, *25*(3), 115-124.

Zhang, Y., Wang, C.L., & Chen, J.Q. (2001). Chinese online consumers' responses to web-based data collection efforts: A comparison with American online consumers. *Journal of Database Marketing, 8*(4), 360-369.

Section II:

E-Marketing Strategy

Chapter V

Global Internet Marketing Strategy:
Framework and Managerial Insights

Gopalkrishnan R. Iyer, Florida Atlantic University, USA

Abstract

This chapter explores some key managerial issues in the development and implementation of a global Internet marketing strategy. While it appears that the Internet has opened up infinite possibilities for an integrated global marketing strategy, this chapter notes several infrastructural and cultural issues that limit the effectiveness of global strategies. This chapter offers insights on using the full potential of the Internet in the deployment of global marketing strategies, while being cognizant of various other realities and limitations. Several practical managerial recommendations are offered for crafting and deploying a global marketing strategy.

Introduction

Even after several years of the commercial expansion of the Internet and the development of the World Wide Web (WWW), the adoption of Internet technologies by global firms and the promises of reaching a global market appear to be fraught with a variety of structural and functional encumbrances (Guillen, 2002; Samiee, 1998). Some of the barriers stem from the relatively slow pace of development of Internet infrastructure around the world, while others are due to the inability of firms to fully exploit the global potential of the WWW. The potential of electronic technologies to foster radical changes to marketing exchanges and organization depends to a large extent on the institutional context of specific firms and countries. Differences in technological advancement, culture, politics, law, and consumer behavior would largely shape the development of electronic markets (Guillen, 2002; Zugelder, Flaherty & Johnson 2000). While such markets offer more complete information as compared to traditional markets, buyer-seller exchanges would, nevertheless, have to adapt to the unique constellation of institutional forces in different countries and industries. Moreover, the regulatory environment of e-commerce is only emerging, and it is bound to be only much more complicated given global differences in intellectual property considerations, patents, consumer privacy, and other issues (see Dutta, Lanvin & Paua, eds. 2003; Kogut, ed. 2003, OECD, 2002; UNCTAD, 2002).

In retrospect, global reach through the Internet may add a new layer of complexity on top of the already numerous challenges faced by firms attempting to cross national boundaries through conventional methods. However, the promise of reaching a wider market more efficiently through the Internet is an opportunity that can certainly be exploited (Brynjolfsson & Kahin, 2000; Gray, 2000; Vulcan, 2003). Savvy firms could harness the powers of the WWW to create superior and customized value for global buyers, search and obtain global resources more effectively, reduce business risk through a diversified portfolio of markets, and thereby, add to profit margins.

This chapter first identifies some strategic and operational opportunities for global marketing presented by the Internet as well as some structural and functional constraints in the realization of the full potential of the Internet for the development of global marketing strategies. Using extant knowledge on global marketing strategy development and implementation, the section following identifies some key issues for the development of a global Internet marketing strategy. This strategy development process is guided by the imperatives of creating a global customer-centric organization. The chapter concludes with some practical managerial guidelines for using the Internet to develop an integrated global marketing strategy.

The Internet and Global Opportunities

Global opportunities presented by the Internet and WWW are both strategic and operational (Samiee, 1988). At the strategic level, some key opportunities are summarized in Figure 1.

Wider Market Access

A key feature of the Internet is the connectivity with diverse and far reaching markets, thus bringing to fruition the type of economic convergence predicted by globalization (Levitt, 1983). Within domestic markets, the Internet can expand the reach and trading areas even for small firms, but the potential for reaching global markets is only exponential in scope (Brynjolfsson & Kahin, 2000; McKnight, Vaaler & Katz, 2001). More firms can chase global dreams and ambitions, and at the very minimum, the ability to attract customers from outside the conventional national trading area more efficiently and effectively enables firms to fulfill their international expansions efforts more rapidly.

Efficient Resource Access

Firms can use the Internet to efficiently search for cross-border resources, including critical raw materials and components, sources of manufacture, and products for which there may be a large domestic demand. With more firms coming online, the amount of trade leads that could be generated over the

Figure 1. Global Internet marketing: Opportunities and challenges

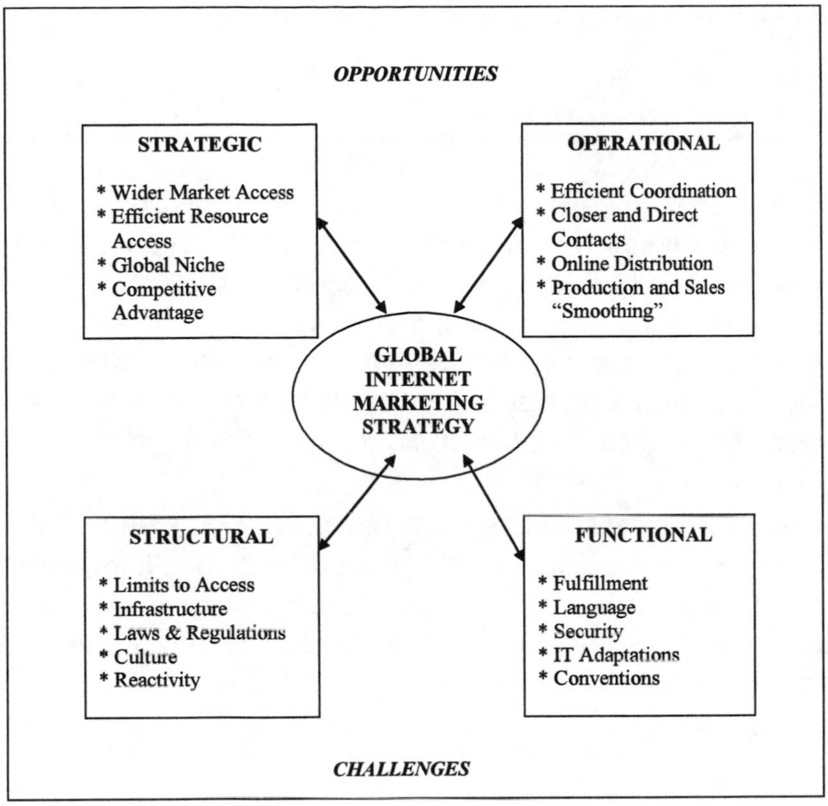

OPPORTUNITIES

STRATEGIC
* Wider Market Access
* Efficient Resource
  Access
* Global Niche
* Competitive
  Advantage

OPERATIONAL
* Efficient Coordination
* Closer and Direct
  Contacts
* Online Distribution
* Production and Sales
  "Smoothing"

**GLOBAL
INTERNET
MARKETING
STRATEGY**

STRUCTURAL
* Limits to Access
* Infrastructure
* Laws & Regulations
* Culture
* Reactivity

FUNCTIONAL
* Fulfillment
* Language
* Security
* IT Adaptations
* Conventions

CHALLENGES

Internet has grown substantially. Electronic tradeboards, business-to-business electronic exchanges, and e-procurement sites are increasingly global in scope, and the search and notification capabilities of such sites can produce a tremendous volume of resource leads for firms.

Global Niche

The Web also enables firms to pursue a specific positioning strategy and applies it globally. For example, Yahoo! has a very similar design for its information and shopping portal in all of the numerous countries in which it has a presence. The portal niche by Yahoo! essentially provides very similar functions for users worldwide, though the content remains local to each country. Contrast this to specific positioning developed and applied painstakingly to each country, and the potential of the Internet becomes more obvious.

Competitive Advantage

Virtual networks formed with participants in the global supply chain, including those involved in production, distribution, and support functions, foster effective coordination and yield tremendous strategic benefits. The new sources of competitive advantage lie in creating and delivering customer value and in the exploitation of network externalities. Ultimately, the collaborative strengths of cross-border alliances become sources of advantage in domestic markets as well. Firms utilizing such global virtual networks could potentially be truly multinational in scope and operations.

The Internet also provides other opportunities to create competitive strengths. Such advantages are no longer location specific but involve knowledge based skills, timing, and flexibility. Firms that do better faster stand to gain more from the Internet, and firms that innovate, create, and apply well-planned business models would better withstand competitive onslaughts.

Some of the operational opportunities provided by the Internet include:

Efficient Coordination

The real-time communication and interactive capabilities of the Internet provide opportunities for much better coordination with suppliers, importers, distributors, and customers than through conventional communication media. But communications are only part of the story. Costs of transactions are lower due to reductions in costs of precontract negotiations and agreements, more efficient price haggling, and more transparent information exchange. Such communication and transactional efficiencies could ultimately contribute to the development of exchanges based more on trust, rather than on costly legalistic frameworks. Commitment enabled through trust is often more enduring and involves lower costs of monitoring as compared to those obtained by fiats and legal documents.

Closer and Direct Contacts

Besides enabling closer contacts with relevant partners and customers, the Internet allows firms to understand them better. Such enhanced understanding could be leveraged for customizing products as well as for better pre- and postsales service design and delivery. Moreover, opportunities for direct sales and global account management are now greatly facilitated, thus reducing dependence on the whims of local agents and dealers.

Online Distribution

For several knowledge and media products as well as information-intensive services, the Internet enables cheaper and efficient online distribution options. For example, software companies could distribute their products online, thus realizing further cost savings. Call centers and technical support could now be located in cross-border markets where labor costs may be cheaper. The increasing media convergence facilitated by the Internet could be used to leverage resources for wider distribution, as traditional news publishing firms and radio stations have already discovered.

Production and Sales "Smoothing"

Variations in demand and supply can be "smoothed" over the global market. The efficient identification of trade leads and resources enable disposal of excess inventory, arranging nontraditional forms of payments (such as barters), and obtaining functional flexibilities, including real-time inventory management and operations scheduling.

Barriers to Global Internet Expansion

Despite the above and obvious global opportunities from the Internet, there are several structural and functional challenges that limit the global reach of firms (Samiee, 1998). Some of the structural barriers include:

Limits to Access

Even among a limited set of developed countries, the U.S. represents more than half of all Web users. Internet access in several countries is currently limited by low levels of PC penetration, high costs of Internet connections, prohibitive software costs and restrictions on use, lack of government initiatives as well as attempts to control both ownership and content by governments, among other factors (Dutta et al., 2003).

Infrastructure

Development of e-commerce requires tremendous investments in infrastructure, including technologies, telecommunications, and human skills – all of which are currently hampered by capital costs, especially in developing countries. For example, while India is very advanced in software development skills and the availability of IT manpower, basic access to the Internet is still limited by the relative low development of telecommunications infrastructure.

Laws and Regulations

The Internet further complicates the already complex conventional laws and regulations governing trade and commerce. While the Internet legal landscape is only evolving, several governments have not yet treated it as a priority issue and appear to simply extend conventional commercial standards to e-commerce. Moreover, issues concerning contracts, arbitration of disputes, as well as laws safeguarding transactions, communications, and intellectual property are yet to be resolved. Attempts to develop unified platforms for such discussions have been more national or, at best, regional in scope, rather than recognizing that the Web has wider trade impacts. The issue of consumer privacy is a case in point. U.S. and European laws regarding consumer privacy are quite different and stem from different historical backgrounds, constitutional rights and limits, and cultural perspectives regarding privacy (Sarathy, 2003). For example, U.S. laws allow a default opt-out policy where consumer information is collected unless the consumer takes active action to prevent data collection and/or use. On the other hand, some European laws are more stringent to business and impose a default opt-in policy. While some convergence in privacy regulations is emerging through agreements, such as the US-EU Safe Harbor provisions for data privacy, firms either subscribe to these agreements or prefer self-regulations through extensive and visible privacy policies (Sarathy, 2003).

Culture

Internet use is also affected by a variety of cultural factors. It is only now being recognized that international consumers may react and receive technology differently, may be unaccustomed to shopping without direct human contact, and may have different predispositions to trust in an online exchange. The implications are clear: business models that work well in one country may not succeed in another. Part of it may be simply that certain shopping habits, such as paying by credit card, which is taken for granted in the U.S., may not be prevalent even in countries like Spain, where credit card usage is less common. Moreover, many international consumers assume that information over the Internet can only be free, as an online test preparation site in India found to its chagrin. When the same material was printed, bound, and sold through traditional distribution channels, consumers willingly paid four times as much as they were being charged online.

Cultural norms and value also affect interpersonal trust, organizational trust, and trust in governments. These differences in trust have profound implications for obtaining trust in one's firm, brands, or building loyalty relationships with customers in different cultures. For example, in countries with low-context cultures, such as the U.S., where most business matters and transactions operate under the *shadow* of detailed and legalistic contracts, trust in a firm's relationships with other organizations and consumers is more readily obtained than in high-context cultures (Hall, 1976). Customer satisfaction in low-context cultures may depend more on rational factors, while customer loyalty may be more emotional. On the other hand, in traditional societies, including even economically developed countries such as Japan, loyalty may be unquestioned. Here, customer satisfaction is often not a rational evaluation but based more on the social context in which the business transaction is embedded. With the Internet and the resulting impersonalization of transactions, firms may find it difficult to obtain customer loyalty in traditional cultures. Cultural norms and values may thus require deeper levels of Web site customization as well as other traditional approaches, such as country presence to signal domestic commitment, in order to build trust and obtain the commitment necessary for relationship development.

Reactivity

Internet use is affected by development of common technological and legal standards, much of which depends on the diffusion of successful world standards across nations. The reaction time could vary, not just from a lack of consensus on basic standards, such as in the case of encryption. Currently, very few countries qualify as ready for e-business.

The functional challenges that limit the operational capabilities of global e-business are also varied:

Fulfillment

It may be easy to obtain an overseas sales order over the Web, but fulfillment is an entirely different issue. Even simple logistics issues, such as transportation and payment, could cause global e-headaches. Currently, no large logistics provider enables a smooth city-to-city link (there are more U.S. city-to-country links), and payment processes are still hampered by lack of trust and elaborate bureaucratic processes designed in principle, though not in effect, to reduce risks. Even the sales order process may be in dispute, as U.S. firms doing business in relatively developed Western Europe are recognizing. Laws protecting small firms may prevent booksellers from offering the same levels of deep discounts in Europe as they do routinely in the U.S. Moreover, physical delivery of the product would be subject to the same trade and nontrade barriers and constraints as any other sales order. Such complications are evident in the case of software distributed online. If the software is classified as a product, it would attract tariff rates and other trade impediments as any other physical product, but if it were to be offered as a service, it can be distributed without much hassle.

Language

Internet-based communication, including Web sites, is constrained by language. Polyglot sites are less common, and the language medium for about 80% of U.S. Web sites is still English (Guillen, 2002). Attempts to customize sites in different languages has only achieved moderate success, partly due to a wholesale use of simplistic translation software and partly also due to lack of recognition, at least by some U.S. firms, that Spanish is quite different when spoken by someone from Bogotá as compared to someone from Barcelona.

Security

Security issues are more complex in global transactions. Country-specific initiatives and regulations for the prevention of hacking, development of adequate security measures, and punishment of offenders are still lacking. However, more importantly, acceptable global guidelines and norms on security and security breaches, so crucial for the development global e-business, are still only talk rather than substance.

IT Adaptations

Several information technology (IT) architectures are still not congenial for global use. At the simplest level, Web site programmers in the U.S. still use the U.S. address and telephone conventions. At a more complex level, there appears little motivation to make significant IT investments to realize e-business opportunities, partly due from viewing IT investments as sunk costs rather than costs involved in opportunity creation and partly due from rapid obsolescence in an era of technological volatility.

Conventions

Operational efficiency is hampered by lack of unified conventions and standards. While several global operations are covered by various ICC codes, such conventions in Internet applications, including the use of digital signatures, are yet to emerge.

Developing Global Internet Marketing Strategy

The development of a global Internet marketing strategy can gain from two of the more popular frameworks of global corporate strategy, namely, the integration-responsive framework proposed by Prahalad and Doz (1987) and the coordination-configuration framework offered by Porter (1986). The integration-responsive framework seeks the global integration of similar activities to achieve efficiency, while ensuring effective local responsiveness to differences across countries. The coordination-configuration framework seeks a balance between the coordination of competitive positions and interdependencies across countries with the configuration of value-adding activities. Both frameworks extol the twin imperatives of efficiency and effectiveness in the organization of global strategy and call for a mix of standardization and adaptation strategies.

Guided by these frameworks on global corporate strategy, a global Internet marketing strategy should also focus on obtaining the best integration of activities that fosters efficiency, while adapting locally to various country-specific environmental contexts. The various structural and operational advantages identified in the previous section contribute to efficiency-enhancing standardization strategies, while the structural and functional encumbrances need effective country-specific strategic adaptations.

The ultimate aim of a global corporate Internet strategy is the creation of a customer-centric global organization that is responsive to the local needs of consumers in a variety of different contexts, while gaining from efficiencies in knowledge management, supply chain integration, technology deployment, and value chain collaborations (Iyer & Bejou, 2003). A framework for such a global strategy is mapped in Figure 2.

Country-specific Internet windows accessible from a single global corporate Web site would provide for value creation, value delivery, and service similar to a locally responsive marketing strategy (Grewal, Iyer, Krishnan & Sharma, 2003). In addition, the technological capability of the Internet would enable customization, personalization, and the required customer analysis for marketing to smaller segments either nationally or within regions of any nation (Iyer, Miyazaki, Grewal & Giordano, 2002).

Back-end processes involving inter-functional coordination, market-oriented logistics, supply-chain management, and technology could be integrated either

Figure 2. The global customer-centric organization

Global Resource Creation	Global Processes	Country-Specific Internet Window	
- Value-Chain Collaborations	- Interdivision Coordination	- Positioning	**Global Customer-Centric Organization**
- Knowledge Development and Deployment	- Market-Based Logistics	- Value Creation	
- Technological Advantages	- Supply-Chain Integration	- Value Delivery	
	- Technological Compatibility	- Service	
		- Personalization	
		- Customization	
		- Privacy Policy	
		- Customer Analysis	

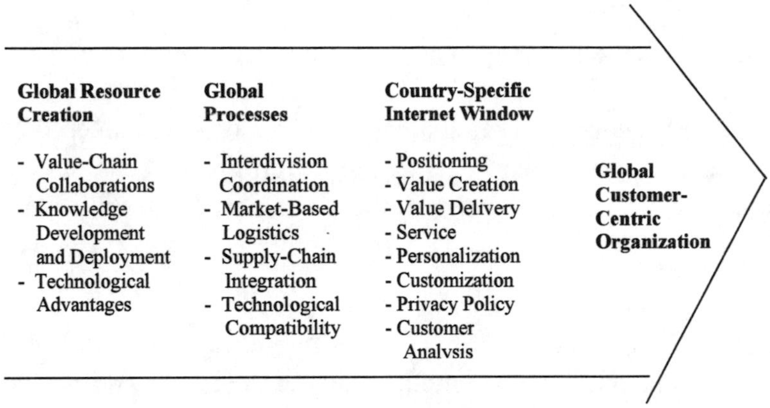

globally or regionally. These, in turn, are facilitated by globally created firm-specific capabilities of the multinational corporation. The multinational firm can harness advantages in terms of relevant value-chain collaborations, knowledge development, and technological deployment (Conn & Yip, 1997). Such capability building and capability exploitation efforts are critical not only for international expansion but also for the ultimate success of the multinational corporation's global strategy (Luo, 1999).

Apart from efficiency and effectiveness, a global Internet marketing strategy based on specific customer needs in each local market enables the firm not only to choose the appropriate value through appropriate target market selection and positioning but also creates additional value for the local customer, and thus, be more competitive in local markets (Grewal et al., 2003; Luo, 1999). The technological capability of the Internet is such as to provide the firm with the ability to create, communicate, and compete along different value propositions in different countries. For example, a consumer electronics MNC may emphasize value and low prices in the U.S. and Europe but emerge as a manufacturer of high quality and upscale products in developing countries.

However, more importantly, firms can offer different product lines, different prices, and communicate and deliver value differently in different countries. A case in point is Hewlett-Packard's global Web site – *www.hp.com*. From the global site, individual consumers as well as business customers can access their specific country's site and be exposed to customized product mixes, prices, and delivery options.

Managerial Implications

The varied challenges in the global e-business environment may dampen the ambitions of many firms. However, the tremendous business opportunities as well as sources of competitive advantage may be sufficient incentives for firms courageous enough to venture into the rough global seas of the Internet. Pragmatic decisions for charting a global course on the Internet could involve the following:

* *Assess the global impacts of the Internet for the firm and industry:* The question is no longer whether the Internet has any relevance for one's firm and industry but how much of an impact it would have on one's specific type of firm and industry. For example, a group of rice farmers in Sri Lanka found the structure of prices for their products and the possible exorbitant markups charged by middlemen through the Internet. Armed with this information, they were able to negotiate better prices for their products at the next round of bargaining with intermediaries.

* *Obtain top management commitment early:* If anything, the lessons learned from conventional cross-border business attempts have been that global expansion strategies rarely work when implemented without a coherent grand strategy and commitment of resources early in the process of internationalization. In global e-business as well, top management commitment obtained early is crucial for support to global decisions regarding IT investments, e-business objectives and strategies, and the integration of e-business activities within the larger organization.

* *What should the Web to do for the firm?:* While the Internet and WWW hold enormous capabilities and extend existing ways of doing business in various markets and industries, an individual firm has to clearly understand the more important uses of the Web for its own specific case. For a vast majority of firms, the Web is simply another form of directory listing, and their sites contain no more than some basic information about the firm. But in all fairness, such sites are often created by small entrepreneurs for whom the costs of developing and maintaining a sophisticated transactions-oriented Web site are much greater than either what their resources allow or the incremental revenues that a WWW presence may yield. However, the global potential of the Web are better exploited in other ways, both for the short-term as well as for a long-term global expansion strategy. In the

short-term or sporadically, firms could benefit from using the Web to search for international suppliers of products and services, importers and exporters for specific products, reduce excess inventory, obtain quotes and bids, and identify prospective international partners and customers. However, the long-term strategy involves a more committed Web presence, with the added costs of designing, developing, and maintaining a secure site. A firm must balance its market objectives, resource needs, competitive capabilities, and product and industry-specific constraints when making this decision.

- *Adapt to country variations:* Despite the lip service given to globalization and the possibility of a world made one through technology, firms must carefully consider the relative variations in technological developments, infrastructural differences, uniqueness of business environments, impacts of culture on business activities, and managerial modes of thinking specific to certain countries/regions. But most importantly, firms must pay close attention to the myriad national laws and regulations, especially those that directly impact e-business activities. Strategy development must be aware of adequate measures and precautions, not only for contracts, but also for IT deployment and information assets. A secure transactional environment is based on several factors, including but not limited to, technological encryptions, legal and enforceable aspects of contracts, and trust.

- *Customize for specific clients:* It should not be a surprise that the Internet is addressable (i.e., identity-based relations can be formed) and therefore, customizable to individual customers and client organizations. It is, however, surprising that few firms use the Internet to customize content and strategies. In global markets, the ability to customize pays rich dividends, especially since local demand-supply conditions and customer expectations vary widely. Customization enables better product development capability, specific service satisfactions, and therefore, greater margins. For example, Dell's site is not only country-specific, but through the creation of Premier Pages, Dell has customized its product and service offerings to its clients. More importantly, the extent to which cultural factors affect Internet use, perceptions of Internet firms and brands, and impersonal Internet transactions should be studied on a case-by-case basis to obtain insights on what aspects of the business can be standardized and which ones need to be customized given the national culture.

- *Adapt the business model:* Regardless of the success of the firm's business model in the domestic market, the chance of similar success in global markets without any adaptations are, at best, slim. Business models can rarely be exported so easily, though some notable exceptions, such as the Yahoo!, have retained a core business model, while creating local content in each country. Take, for example, auction sites. It may sound intuitive that auction-based pricing would be tremendously successful in the traditional countries of Asia and the Middle East, where price haggling is more of a norm in market transactions. However, auctions sites, such as eBay and *QXL.com* have been more successful in the U.S. and Europe, respectively, where fixed prices are more common. Contrast this to India, where a take-off of eBay, *Bazee.com*, suffered from being perceived as a bankruptcy liquidating site. The situation is more complicated for B2B sites, especially since most intermediary organizations–suppliers, distributors, importers, exporters, retailers–outside of a few developed countries, are generally much less sophisticated in the use of technology. They may also be much less professional, preferring old-world identity and reputation-based ties to rational market-based exchanges.

- *Focus on brand-building*: It has been long recognized that a key factor in domestic e-business success has been the success of the Internet brand. Internet-only firms, such as Amazon.com and Yahoo!, were able to leverage their pioneering advantages and build their reputations through online and traditional media. While such pioneering opportunities may not exist around the world, successful global brands can still be built by focusing on the factors contributing to the firm's reputation. Here, the power of global consumer associations, tangible consumer values from the brand, pre- and postsales services provided by the firm, the firm's responsiveness to consumers, customization and interactivity of the Web site, and the consonance of the brand with consumer values all play important roles. Firms must carefully consider all aspects of consumer value from the brand, consumer demands for service, and consumer proclivity to identify with the firm's brand, and factors that contribute to positive word-of-mouth or word-of-mouse. While large firms can leverage their existing brands into global brands over the Internet, smaller and resource-constrained firms could focus on specific product level brand reputations. Providing a superior value and careful positioning of the brand are important and so are providing accurate information through the Web site, superior service (including critical information and handling complaints) and keeping promises.

- *Develop capabilities:* The various challenges in enabling a global e-business can be better countered with external help. There are by now hundreds of tradeboards, several dozen vertical marketplaces and B2B hubs that include global buyers and suppliers, and probably several hundred firms offering specific local country expertise in Web design, language translations, and transactional and operational logistics as well as shipping alternatives. Outsourcing various elements of the global e-business strategy helps obtain better local skills and ensures a low cost exit strategy, while reducing the risks involved in making large in-house IT investments. Partnering may be a quick way to gain competence and also search for unique and innovative solutions (Iyer, 2004).

- *Innovate, adapt, and improvise:* A successful global e-business strategy hinges on (1) innovating to counter global e-challenges and exploit global e-opportunities; (2) adapting business models, content, and transactions to local environments; and (3) the willingness to improvise swiftly based on available facts. Innovations in solutions for the design and delivery of products and services go a long way in exploiting the market creation potential of the Internet. Adaptations to local environments minimize the risks of failure and enhance local competitiveness, while fast-paced improvisations enable the firm to capitalize on new opportunities in the global market.

Suggestions for Future Research

The preceding discussions on the opportunities and challenges in global Internet marketing as well as the managerial implications suggest several areas where further research may be needed.

Comparative Studies on Internet Marketing

While several international organizations have started measuring and comparing the relative levels of Internet infrastructure and e-commerce development around the world, specific research is definitely needed on the comparative similarities and differences in the conception, development, and practice of Internet marketing in different parts of the world. Research is needed on studies that uncover consumer perceptions and use of the Internet for shopping, the relative levels of development of B2B business infrastructure and practices, the impacts of various Internet-related and firm strategies on e-business success and marketing performance, and the extent to which customer relationship management when practiced over the Internet differs across various countries and cultures.

Impacts of Institutional Factors on Internet Marketing Success

The preceding discussion has only scratched the surface on the myriad institutional factors, including but not limited to, laws and regulations and cultural institutions that must be considered when developing a global Internet marketing strategy. However, the specific impacts of such institutions on the development and implementation of Internet marketing strategies need to be the foci of future research. Research could also uncover how historic institutions affect the practice and success of Internet marketing around the world and how such institutions contribute to marketing and operational efficiencies over the Internet.

Customer-Centricity and Relationship Management

Research is needed on how the relationship process (i.e., the links between customer satisfaction, loyalty, and profitability) differs across various countries and cultures and how the Internet enables or hampers the development of profitable customer relationships. The preceding discussion has contended that the presumed linkages may not exist uniformly across the world, but only directed future research can reveal the exact nature of the similarities and differences in relationship processes and the comparative efficacy of relationship management practices.

In conclusion, the Internet is a gold mine of global business opportunities, but the ability to find the mother lode depends upon precise knowledge of the terrain, development and use of sophisticated and innovative tools to overcome barriers, leveraging the expertise of significant partners, pace of competitive capability building, and, of course, serendipity.

Acknowledgment

The author thanks the Lynn Chair Research Grant and the InternetCoast Institute Adams Professorship for research support.

References

Brynjolfsson, E., & Kahin, B. (eds.). (2000). *Understanding the digital economy: Data, tools, and research*. Cambridge, MA: MIT Press.

Conn, H.P., & Yip, G.S. (1997). Global transfer of critical capabilities. *Business Horizons, 40*(1), 22-32.

Dutta, S., Lanvin, B., & Paua, F. (eds.). (2003). *The global information technology report 2002-2003*. New York: Oxford University Press.

Gray, R. (2000, April 13). Make the most of local differences. *Marketing*, 27-28.

Grewal, D., Iyer, G.R., Krishnan, R., & Sharma, A. (2003). The Internet and the price-value-loyalty chain. *Journal of Business Research, 56*(5), 391-398.

Guillen, M.F. (2002, May-June). What is the best global strategy for the internet? *Business Horizons*, 39-46.

Hall, E.T. (1976). *Beyond culture*. New York: Anchor Books/Doubleday.

Iyer, G.R. (2004). Internet-enabled linkages: Balancing strategic considerations with operational efficiencies in B-to-B marketing. *Journal of Business-to-Business Marketing, 11*(1/2), 35-59.

Iyer, G.R., & Bejou, D. (2003). Customer relationship management in electronic markets. *Journal of Relationship Marketing, 2*(3/4), 1-17.

Iyer, G.R., Miyazaki, A.D., Grewal, D., & Giordano, M. (2002). Linking Web-based segmentation to pricing tactics. *Journal of Product and Brand Management, 11*(5), 288-300.

Kogut, B. (ed.). (2003). *The global internet economy*. Cambridge, MA: MIT Press.

Levitt, T.H. (1983, May-June). The globalization of markets. *Harvard Business Review, 61,* 92-101.

Luo, Y. (1999). *Entry and cooperative strategies in international business expansion*. Westport, CT: Quorum Books.

McKnight, L.W., Vaaler, P.M., & Katz, R.L. (Eds.). (2001). *Creative destruction: Business survival strategies in the global Internet economy*. Cambridge, MA: MIT Press.

OECD (2002). *OECD information technology outlook: ICTs and the information economy*. Paris: OECD.

Porter, M.E. (1986). Changing patterns of international competition. *California Management Review, 28*(Winter), 9-31.

Prahalad, C.K., & Doz, Y.L. (1987). *The multinational mission: Balancing local demands and global vision*. New York: Free Press.

Samiee, S. (1998). The Internet and international marketing: Is there a fit? *Journal of Interactive Marketing, 12*(Autumn), 5-21.

Sarathy, R. (2003). Privacy protection and global marketing: Balancing consumer and corporate interests. In S. Jain (Ed.), *Handbook of research in international marketing* (pp. 358-376). Cheltenham, UK: Edward Elgar.

UNCTAD. (2002). *E-commerce and development report 2002*. New York/Geneva: Author.

Vulcan, N. (2003). *The economics of e-commerce: A strategic guide to understanding and designing the online marketplace*. Princeton, NJ: Princeton University Press.

Zugelder, M.T., Flaherty, T.B., & Johnson, J.P. (2000). Legal issues associated with international internet marketing. *International Marketing Review, 17*(3), 253-271.

Chapter VI

Interactive Brand Experience:
The Concept and the Challenges

Mary Lou Roberts, University of Massachusetts Boston, USA

Abstract

E-marketing has evolved from an environment that was solely direct response to one that includes significant opportunities for brand development. Internet marketing uses a rich set of tools to create exceptional customer experience that results in strong relationships with the brand. Off-line tools like elements of the brand and image-building promotional programs are as important as ever in building brand equity.

To those have been added online tools, including personalization, customization, cocreation, purchase-process streamlining, self-service, brand community, rich media, product self-design tools, and dynamic pricing. This gives the marketer a rich choiceboard of off-line and online techniques from which to select the most appropriate for the brand and any given promotional activity. The challenges are to select the most effective tools and then to execute flawlessly and seamlessly at all customer touchpoints to achieve the desired quality of customer experience.

Introduction

From the onset, marketers have regarded the Internet as the consummate direct-response medium. The ability to interact one-on-one with customers and the ability to track their every move allowed precision targeting never before possible. More recently, it has become clear that the Internet can also be used in branding efforts. The ability to blend direct response and branding efforts is, at the same time, the Internet's greatest benefit and its ultimate challenge to marketers. This chapter will examine online branding techniques and consider their effectiveness. It will then propose a branding concept that will help to integrate marketing activities in both interactive and traditional media environments and discuss the challenge posed by the need to integrate communications channels. Since the acceptance of the Internet as a venue for brand development is a relatively new phenomenon, even in terms of the existence of the Internet, we begin there.

Branding Efforts on the Internet

A direct marketing mindset permeated successful Internet marketing activities in the early years, one that valued immediate and measurable response and had little patience for marketing effort with long-term payoffs. It is still reflected today in the majority of the metrics–hits, visitors, page views, for example–that are the stock in trade of measurement on the Web. As early as 1997, Briggs and Hollis challenged the widely-held belief that click-through on banner advertisements was the most important measure of Internet success. Their experiment found an increase in favorable attitude in all three product categories used in the study. The impact was greatest in a brand never before advertised on the Web. The most comprehensive and best-known study of branding effort on the Internet is the Cross Media Optimization Study of the Interactive Advertising Bureau. Begun in 2002, the study includes over 30 leading brand marketers and on- and off-line publishers as participants. Methodology builds on established off-line metrics by adding accepted online measures.[1] Selected studies provide evidence that Internet advertising does affect various brand metrics.

- One of the earliest was conducted in conjunction with the introduction of Unilever's Dove Nutrium brand. The basic research design was to run print advertising only in week one, add online in week two, and television in week 3. When cost of media placements were included, the study concluded that keeping the total advertising budget constant but increasing online spending from 2% to 15% would produce an 8% increase in overall branding impact and 14% increase in purchase intent (*http:// www.iab.net/xmos/pdf/xmosdatadove.pdf*, no date).

- Another study focused on Kimberly-Clark's introduction of the Kleenex Soft Pack. The media allocation was 75% to television, 23% to print, and 2% to online. The study found that online advertising reached the 42% of the target audience that is not reached or only lightly reached by television (*http://www.iab.net/xmos/ pdf/ xmosdatadove.pdf*, no date).

- A recent study for Volvo used the Sponsorship Effectiveness Index to compare the effectiveness of shared sponsorship (multiple ad placement on a single Web page) with exclusive sponsorship in which no other advertising is present on that particular page at that particular time. The study concluded that shared exposure resulted in no significant lift in brand inclusion in the consideration set while exclusive sponsorship resulted in a 6.1% increase in brand inclusion in the consideration set (*http:// www.iab.net/resources/iab_volvo.asp*, 2003).

Other organizations report similar results. British marketing research firm Taylor Nelson Sofres Interactive conducted four separate studies during 2000 and 2001. The studies showed that online advertising generally did increase brand awareness, more for unfamiliar and less for familiar brands. However, higher levels of ad recall were not always correlated with higher levels of brand awareness (Hughes, 2002). A 2003 study by agency *Advertising.com* monitored conversions from a credit card offer over a five-day period. They found that about 33% of the conversions occurred on the same day as ad exposure, but only 11% occurred within three hours. In another study, when viewer activity was monitored for 14 days after initial impression, as many as 85% of the conversions occurred more than one day after exposure (Advertising.com, 2003).

At this point, the evidence seems sufficiently strong to accept the Internet as a branding medium, the premise that will drive the remainder of this chapter. That permits us to examine the ways in which the medium can be best used in branding campaigns. In order to do so, we must first examine the major brand development models.

Building and Maintaining Brands

In recent years, two models of brand development have emerged. They are complementary, not competitive.

Brand Equity

Arguably the most widely accepted brand development model is Keller's Customer-Based Brand Equity Framework (Keller, 1998). It is composed of tools and objectives (brand elements, marketing programs, and secondary brand associations) that are mediated through knowledge effects (brand awareness and associations) with resulting enhancements of brand equity that include larger margins and greater brand loyalty. Keller expanded on the static model by providing a series of steps for creating a strong brand: establish the proper identity, create the appropriate brand meaning, elicit the right brand responses from customers, and forge strong relationships with them (Keller, 2001).

Ilfeld and Winer (2002) studied the development of brand equity on the Internet. They used a traditional hierarchical approach, adapted to take Internet differences into account. This allowed them to test three models: persuasive hierarchy (Think-Feel-Do), low-involvement (Think-Do-Feel), and no involvement (Do-Think-Feel). Overall, the Think-Do-Feel model performed significantly better on all measures, suggesting that awareness is followed by site visitation, which, in turn, is followed by brand equity. They liken Web visitation (the dependent variable) to mature frequently purchased product categories (low involvement) in which advertising is useful in building awareness and driving usage and note that both online and off-line efforts are required.

Brand Relationship

A different approach is taken by Fournier (1998). In a study of how consumers develop relationships with their brands, she advances the concept of brand relationship quality (BRQ). BRQ is a multidimensional construct composed of positive affective feelings (love/passion, self-connection), behavioral ties (interdependence, commitment), and cognitions (intimacy and brand partner quality). BRQ is mediated by a number of psychosocial filters with the outcome determining the stability and durability of the consumer/brand relationship.

Thorbjørnsen et al. (2002) operationalized the BRQ dimensions and tested whether customer communities (person-to-person interaction) or personalized Web sites (machine-to-person interaction) were most effective in building BRQ for hypothetical products. They found that for less experienced consumers, customer communities were more effective. For more experienced users, personalized Web sites were more effective.

Fournier's (1998) model specifies an outcome–the quality of the relationship that consumers have with their brands. It is an outcome predicated both on consumers' own life experiences and brand-related marketing actions. Marketers cannot control consumers' life experiences; however, they can create and exercise control over customer experience. We therefore turn next to the concept of consumer experience.

Customer Experience

The concept of customer experience as a key to brand learning predates the Internet. Hoch and Deighton (1989) characterize it as a type of learning, a 4-stage information processing model. They postulate that consumers formulate working hypotheses for testing, are exposed to evidence about the product, encode information based on their own familiarity and motivations, and finally integrate new evidence into their existing belief structures. They distinguish between learning by description (most advertising falls into this category) and learning from experience, which is recognized as more effective. Pine and

Gilmore (1999) popularized the concept of customer experience, saying that "companies stage an experience whenever they engage customers, connecting with them in a personal, memorable way" (p. 3). The word *stage* is important and reflects Hoch and Deighton's contention that marketers can control the experiential learning of consumers.

Li, Daugherty, and Biocca (2001) argue that virtual experience is similar to indirect experience in that it is mediated. It is also similar to direct experience in that both are interactive. Their research found virtual experience to be an active psychological process. It was accompanied by three other characteristics: presence–which they define as a sense of being in another place generated by indirect experience–involvement, and enjoyment. They conclude that virtual experience consists of vivid, involving, active, and affective states that are closer to direct than to indirect experience.

The concept of customer experience encompasses all the marketer-initiated activities that influence brand equity. Learning, or indirect experience, takes place in the traditional media. Direct experience is gained at the point of purchase and in actual use. Experience gained through interactive media, particularly the Internet, has characteristics of both direct and indirect experience. Figure 1 summarizes the relationship of the brand development concepts. It identifies customer experience as the mediating factor between the marketer's efforts to create brand equity and the consumer's perception of the quality of relationship with the brand. It is therefore critical that marketers use all the techniques at their disposal to create meaningful experience, a subject to which we will return at the end of this chapter. The Internet is the newest of the channels for experience creation, and it is likely that developments in technology will generate even more branding techniques in the years to come. It is

Figure 1. The role of customer experience in brand development

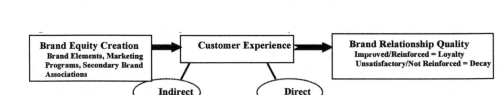

critically important that marketers understand the existing techniques, integrate those into their existing marketing programs, and be vigilant as new techniques continue to emerge.

Online Brand Development Techniques

In 2000, consultants at McKinsey identified tools that can be used to create digital brands–personalization, customization, cocreation, purchase-process streamlining, self-service, product design toolkits, and dynamic pricing (Dayal, Landesberg & Zeisser, 2000). Since then, technology has added rich media to the list. Research has also demonstrated that brand community can be built or strengthened on the Web (see, for example, Bagozzi & Dholakia, 2002). Astute use of these tools and techniques creates an *interactive* environment in which meaningful dialog can take place between marketer and customer. Each of these techniques brings interesting possibilities to branding programs. It is also important to keep in mind that these techniques are available in addition to the traditional brand elements and secondary associations. Elements like the brand name and logo and associations like the company and its channels of distribution are the foundation of all branding activities. Before we look at the specific interactive techniques, we should consider the concept and importance of interactivity itself.

Interactivity

In an early and influential study, Hoffman and Novak (1996) introduced the concept of the computer mediated environment (networks) in which both machines and persons could interact in a one-to-one or many-to-many fashion. There are still interactive media that do not directly involve the Internet–CD-ROMs and ATMs come to mind–but higher access speeds have made the Internet the focal point of interest in interactivity. Hoffman and Novak (1996) also introduced the concept of flow to the marketing literature. They describe it as a state in which "irrelevant thoughts and perceptions are screened out and the consumer focuses entirely on the interaction" (p. 58).

It seems reasonable to assume that brand-related learning is enhanced by the flow state, but there is no evidence that flow is required. Liu and Shrum (2002) did, in fact, find that higher interactivity provided a more involving experience that lead to greater user learning. Their definition of interactivity includes three dimensions: active control by the user, two-way communication, and synchronicity (simultaneous input and response). A later study by Liu (2003) developed an interactivity scale using the three dimensions that were shown to have both validity and reliability. An empirical study of interactivity by McMillan and Hwang (2002) revealed three dimensions that are similar but not precisely the same. They called their dimensions real-time conversation, no delay, and engaging.

These studies support the conventional marketing wisdom (see, for example, Underscore Marketing, 2003) that asserts that well-done interaction engages the visitor, enhancing learning and increasing the likelihood of immediate or delayed behavioral response. This, in turn, suggests that the concept of interactive customer experience is worth pursuing. Figure 2 shows the interactive brand development tools along with an assessment of whether they are primarily the result of machine interactivity or person interactivity or whether they can be either, depending on the nature of the marketing program employed at a given time. We will briefly examine the empirical evidence supporting use of the tools. The evidence is useful, but it does not provide a road map for the practicing manager who wishes to use the techniques. Consequently, recommended executions will be suggested. These are stated as managerial propositions, but they could be reframed as research propositions.

Personalization

Database marketers have personalized communications with name and targeted content for at least two decades. It is a technique easily transferred to the Web. Applications range from greeting an identified user by name when entering a site to using active server pages to deliver content targeted to the user's known interests to serving advertising on the basis of anonymous profiles or profiles of registered users.

Marketers appear to believe that personalization is good and that more personalization must be even better. Witness the training of service personnel that instructs them to greet customers by name. Clearly, there are some environments in which personalized customer contact produces a profitable return. Harrah's Entertainment Inc. has invested heavily in loyalty programs and databases to identify and profile its high-value customers. Salespeople (direct experience) have data at hand that show customer value and allow them to offer incentives, such as better rooms to entice high-value customers. Their system also drives impersonal contact (indirect experience), including follow-up mailings to invite customers to return to Harrah's (Levinson, 2001). Payback on personalization efforts are not always realized, however. A study by Jupiter Communications did not find that personalization of Web sites influenced consumers to purchase more. Easy navigation was, however, found to be effective. The study also points out that personalization of e-mails does generate greater impact (Gonsalves, 2003). Holland and Baker (2001) suggest that personalization is effective for both task-oriented and experiential Web site users if it permits them to better fulfill their objectives.

Dahlen, Rasch, and Rosengren (2003) found that visits to sites of expressive high-involvement products were longer with more pages visited and resulted in more positive brand attitude while visits to sites of functional low-involvement products produced less activity and no change in brand attitude. These studies suggest that marketers should carefully consider the nature of the product and the motivation of Web site visitors before investing in personalization.

Recommended execution. Personalization that seems forced or irrelevant–for example, greeting the visitor by name when entering a site–provides little benefit. However, personalization that allows the visitor to complete a task quickly and efficiently will be welcomed. This type of personalization ranges from customized site content to segmented e-mails with variable content.

Customization

It seems reasonable that the next step after personalization of content would be customization of products and services. There is, however, confusion between definitions of personalization and customization that renders this statement hard to evaluate. Nunes and Kambil (2001) define personalization as reliance on algorithms that uncover patterns in customer data and extrapolate to make recommendations directly to consumers or to personalize Web page content. They contend that customization lets site visitors specify the desired content. Wind and Rangaswamy (2001) carefully use the term *mass customization* to refer to products that are made to customers' specifications using flexible manufacturing techniques. They add a new concept, customerization, which they define as transforming marketing from seller-centric to customer-centric. Ansari and Mela (2003) call the process of providing individualized content *e-customization.* Their research indicated that optimization techniques that varied both content and order of presentation in persuasive e-mails could increase the click-through rate.

It seems reasonable to employ the term *mass customization* to refer to made-to-order products, as practiced by some manufacturers for over a decade (Pine, Victor & Boynton, 1993). Then personalization can be used to describe marketer-initiated individualization (either machine-to-machine or person-to-machine), and customization can be reserved for customer-initiated choice of content alternatives in computer-mediated environments. Personalization and customization then become clearly differentiated alternatives for the marketer to employ in pursuit of satisfying interactive customer experiences.

Recommended execution. Customization of content from choice menus–for example, personalized pages on portals and content sites–is widespread. It transfers control to the customer who selects either site or e-mail content, and customers welcome this type of control. Which products are worth the extra effort of customization is a matter for marketer testing. Expensive bicycles have been customized for several years, and Land's End's offering of custom jeans and chinos has expanded to blouses and shirts, suggesting success. The difficulty of finding exactly the right fit in a particular product, regardless of its price, may be the relevant variable.

Figure 2. The interactive branding tools

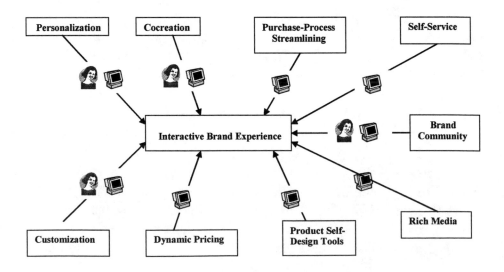

Cocreation

Sawhney and Prandelli (2000) called the phenomenon *communities of creation* in which knowledge is created in a distributed network. Prahalad and Ramaswamy (2000) applied the concept to marketing when they pointed out that consumers have become active players in creating content through focused dialog, software tools such as collaborative filtering, and online communities. They state that customers can help to cocreate their own experiences and, in so doing, determine the level of involvement they wish to have with the brand. Whether the process involves direct personal contact as occurs in eBay auctions, or is machine-based like Amazon's use of collaborative filtering, cocreation can be a highly engaging activity.

Recommended execution. Customers like to converse and contribute in a variety of environments ranging from Amazon to B2B procurement sites. Marketers should use collaborative techniques not only to create content but also to gain useful data and insight about customers' behaviors and attitudes.

Dynamic Pricing

Priceline in the consumer marketplace and FreeMarkets in B2B provide evidence that dynamic pricing can be a key element of an Internet business model. Wyner (2001) points out that customers experience price in a different way on the Web. In the physical world, price is most often a take-it-or-leave-it feature. On the Internet, consumers have a range of options they can use to search for the best prices and decide which e-retailer to patronize. They begin to consider price as a transparent and reasonably controllable product feature.

Recommended execution. Marketers must recognize and try to take advantage of the availability of extensive price information on the Internet. For low cost/low price marketer, this may be relatively easy. For marketers who cannot offer lowest prices, turning to a differentiation technique like product self-design may be an option.

Product Self-Design Tools

A choiceboard can be used to implement product self-design. Bovet and Martha (2000) state that a choiceboard has four key elements: a communication process, the ability to capture real demand, real-time choice management, and the resulting configured-to-order product. Thomke and von Hippel (2002) describe the process by which some business marketers are moving aspects of product innovation activity to their customers. They use the term *tool kits* to describe rapid-prototyping methods that are being deployed for customer use. Whatever the term used to describe the methodology, experts agree that customer product self-design is likely to experience rapid expansion in the coming years.

Recommended execution. As in customization, the customer's desire for control over product features may be a powerful force in many categories. Marketers should ensure that the options being offered are relevant and that the self-design tools work seamlessly. A customer-unfriendly experience may discourage potential users for some time to come.

Purchase-Process Streamlining

The best-known example of purchase-process streamlining is Amazon's 1-Click process. Their customers can enter information about purchaser and payment method, which is then stored for future use. It is also the most controversial example because Amazon applied for and received a patent for the underlying software that has been disputed by other Internet retailers (Gleick, 2002). Nevertheless, other sites have implemented processes that make it easy to purchase, increasing the likelihood that customers will return to those sites for future purchases.

Recommended execution. Filling out the required e-commerce forms can be an annoying task, especially if they do not work properly at that particular moment. Consequently, purchase-process streamlining should be used whenever possible. However, saving data that includes customer credit card numbers puts a heavy burden of security on the marketer. Marketers can use this to their advantage by creating a secure purchasing environment and communicating this to customers in a manner that fosters trust.

Self-Service

Self-service, either on the Internet or through other electronic technologies, is a growing force as companies try to provide customers with more service alternatives at the same they lower their service costs. Meuter et al. (2000)

found 56% of their respondents describing satisfactory self-service encounters and 44% reporting unsatisfactory encounters over a broad range of technologies. Moon and Frei (2000) suggested that, in order to make self-service satisfactory, companies must limit the range of choices and streamline the self-service process. In a study of the effect of technology anxiety on use of self-service options, Meuter et al. (2003) identified five benefits of self-service: effectiveness, enjoyment/independence, higher quality, cost savings, and lack of other options. Howard and Worboys (2003) add that firms must actively promote the benefits of self-service, which they identify primarily as cost and time saving, in order to promote consumer adoption. Overall, it appears that consumers are willing to use self-service options when they solve a difficult problem, are better than the alternatives, and work properly (Bitner, Ostrom & Meuter, 2002).

Recommended execution. Transitioning Internet customers from more expensive service options like call centers to self-service is in the economic best interests of marketers. However, they need to ensure that the service options are robust and that they include human interaction if self-service cannot resolve the customer's issue.

Rich Media

While there is no universally accepted definition of rich media, the IAB defines it as "a method of communication that incorporates animation, sound, video, and/or interactivity" (*http://www.iab.net/resources/glossary_r.asp*). Ad serving firm DoubleClick (2003) reported that rich media increased from 17.3% of ads served in the first quarter of 2002 to 36.6% in the third quarter of 2003. Further, they report that for firms using the click-through metric, click-throughs were greater from rich media ads than from traditional banner ads (1.57% to .29%) (*http://www.doubleclick.com/us/knowledge_central/documents/trend_reports/dc_q303adservingtrends_0310.pdf*). Marketing research firm Dynamic Logic (2002) finds that rich media ads are twice as effective as traditional ads in lifting the message association metric (the ability to link a specific brand with a specific message).

Recommended execution. The empirical data suggest that rich media can be a powerful marketing tool for both direct response and branding. If intrusive–for example, pop-ups–it can also annoy consumers. Marketers therefore need to test rich media for appropriate offers and carefully measure the lift it provides to their marketing efforts as well as any backlash it creates among customers.

Brand Community

Defined as "a structured set of social relationships among admirers of a brand" (Muniz & O'Guinn, 2001), brand community is not an Internet-specific phenomenon. The H.O.G. (Harley Owner's Group) association of motorcycle enthusiasts has been extensively documented (see, for example, Fournier, McAlexander & Schouten, 2000). McAlexander, Schouten and Koenig (2002) studied the Jeep user group. Muniz and O'Guinn (2001) studied Saab and MacIntosh user groups, noting the importance of product-related Web sites in group activities. Holland and Baker (2001) note that many Web sites are encouraging the development of virtual communities in the belief that they promote brand loyalty. Bagozzi and Dholakia (2002) add that consumers are motivated to participate in virtual communities both by personal and group intentions, expecting to derive satisfaction on both dimensions.

Recommended execution. Marketers are looking for features that will draw users to their sites more often for longer visits, and brand communities may be one answer. They are, in fact, one type of cocreation of content and should be treated as both a customer-pleasing feature and a source of marketing-relevant data.

This overview of Internet brand development tools has made it clear that many are not, in fact, unique to the Internet. They are also practiced in the off-line world by direct personal or indirect media contact. They are all, however, well suited to the interactive environment of the Internet and therefore represent an especially valuable set of marketer options.

The Marketing Challenge

From the beginning of this chapter, it has been stressed that Internet brand development techniques do not exist in isolation from the traditional off-line methodologies. The challenge for the marketer is twofold. First is to identify the appropriate techniques and the media best suited to deliver them. Second is to execute seamlessly at all contact points.

These contact points (often called *touchpoints* in the trade literature) include human, impersonal, and technological channels. The Internet, other electronic media, retail stores, and events are characterized as producing direct brand experience because of their interactivity and immediacy. Using the same criteria, sales promotions, mass media, and public relations produce indirect brand experience. Traditional direct marketing, which is interactive but lacks immediacy, can be thought of as producing indirect brand experience.

Many, if not most, marketers today use multiple channels of distribution as well as multiple communications channels. Retail stores, catalogs, and a Web site represent one typical set of channels. Appropriate communications channels must be chosen for each specific marketing initiative. Interactive user-driven experiences are appropriate for the Web while events can be used in either the retail or Internet channels. Catalogs are a traditional direct marketing technique. And so it goes, with an extensive set of combinations available for use. The challenge to marketers is complicated even further by the variety of technologies that can be employed in the communications process. A retail bank can offer its customers the ability to interact through retail branches, telephone call centers, ATMs, or over the Internet. Many are now considering adding wireless Internet banking to the existing technological infrastructure.

Marketers, then, must choose distribution channels and communicate through them in appropriate ways using the most effective technologies. The number of possible combinations is enormous. The difficult of simultaneously executing with excellence in all of them is marketing's great contemporary challenge.

Managerial Implications

There are a number of factors that marketers should consider as they confront this marketing strategy challenge. The first is that it is a moving target. The marketing choiceboard discussed in the preceding section is already complex; technology is likely to add more options as time goes on. It is important that managers stay abreast of changing technology and adopt it when it provides value. It is equally important to reject technology for technology's sake.

Marketers are only in the beginning stages of mounting complex interactive programs that involve the Internet. There is no road map to guide the many choices enumerated in this chapter. Research is important. Testing of programs, as practiced in direct marketing for many years, is essential.

Research indicates that most Internet users are task focused, but there may also be segments that are entertainment- or experience-oriented.[2] Marketers need to facilitate task accomplishment in all possible ways, many of which are represented by the Interactive branding tools. The experience-oriented consumer is eager for interactivity, often represented by *advertainment*, *edutainment*, or traditional games. The objective must be to embed the brand message in experience-producing activities. Internet metrics like return visits and time spent on site are useful in assessing how well customers are being served. Relevant metrics are also available for advertising and e-mail programs. Development of customer scenarios can also help marketers understand how consumers use their site and their electronic communications. The empirical evidence about Interactive Brand Experience and the tools used to create it presented in this chapter show that we are beginning to accumulate valuable data about individual tools. The research to date, however, focuses almost exclusively on individual tools in individual channels. Except for the IAB studies, there is virtually no published research on the effect on brand development of executions of the same message in various channels or the use of different supporting messages in diverse channels. Understanding interactions between channels and messages, both online and off-line, is critical to success in multichannel marketing. Increased understanding of the interactions will help marketers meet the challenge of using the right combination of tools to create Interactive Brand Experience that support and enhance overall customer experience and foster the development of strong brand equity.

References

Advertising.com. (2003). View-through performance. Retrieved October 12, 2004, from *http://www.advertising.com/Press/03Oct28.html*

Ansari, A., & Mela, C. (2003). E-customization. *Journal of Marketing Research, XL*, 131-145.

Bagozzi, R.P., & Dholakia, U.M. (2002). Intentional social action in virtual communities. *Journal of Interactive Marketing, 16*, 2-21.

Bitner, M.J., Ostrom, A.L., & Meuter, M.L. (2002). Implementing successful self-service technologies. *Academy of Management Executive, 16*, 96-109.

Bovet, D., & Martha, J. (2000). *Value nets*. New York: John Wiley & Sons.

Briggs, R., & Hollis, N. (1997). Advertising on the Web: Is there response before click-through? *Journal of Advertising Research, 37*, 33-45.

Dahlen, M., Rasch, A., & Rosengren, S. (2003, March). Love at first site? *Journal of Advertising Research,* 25-33.

Dayal, S., Landesberg, H., & Zeisser, M. (2000). Building digital brands. *McKinsey Quarterly, 2.* Retrieved October 12, 2004, from *http://www.mckinseyquarterly.com*

DoubleClick. (2003). DoubleClick Q3 2003 ad serving trends. Retrieved October 12, 2004, from *http://www.doubleclick.net*

Dynamic Logic. (2002). Rich media campaigns twice as effective at lifting brand message association. Retrieved October 12, 2003, from *http://www.dynamiclogic.com*

Fournier, S. (1998). Consumers and their brands: Developing relationship theory in consumer research. *Journal of Consumer Research, 24,* 343-372.

Fournier, S., McAlexander, J., & Schouten, J. (2000). Building brand community on the Harley-Davidson posse ride. *Harvard Business School, 9,* 501-015.

Gleick, J. (2002). Patently absurd. Retrieved October 12, 2004, from *http://www.nytimes.com.*

Gonsalves, A. (2003). Personalization said to be ineffective on online retail sites. *TechWeb.* Retrieved October 12, 2004, from *http://www.Internetweek.com/story/showArticle.jhtml?articleID=15300234*

Hoch, S., & Deighton, J. (1989). Managing what consumers learn from experience. *Journal of Marketing, 53,* 1-20.

Hoffman, D.L., & Novak, T.P. (1996). Marketing in hypermedia computer-mediated environments: Conceptual foundations. *Journal of Marketing, 60,* 50-68.

Holland, J., & Baker, S.M. (2001). Customer participating in creating site brand loyalty. *Journal of Interactive Marketing, 15,* 34-45.

Howard, J., & Worboys, C. (2003). Self-service – A contradiction in terms or customer-led   choice? *Journal of Consumer Behavior, 2,* 382-392.

Hughes, G. (2002). The branding power of advertising online: Theory & evidence. Retrieved October 12, 2004, from *http://www.tnsofres.com/industryexpertise/Internet/brandingpoweronline.pdf.*

Ilfeld, J., & Winer, R. (2002). Generating website traffic. *Journal of Advertising Research, 42,* 49-61.

Keller, K. (1998). *Strategic brand management..* Upper Saddle River, NJ: Prentice Hall.

Keller, K. (2001, July-August). Building customer-based brand equity. *Marketing Management,* 15-19.

Levinson, M. (2001, May). Harrah's knows what you did last night. *Darwin.* Retrieved October 12, 2004, from *http://www.darwinmag.com*

Li, H., Daugherty, T., & Biocca, F. (2001). Characteristics of virtual experience in   electronic commerce: A protocol analysis. *Journal of Interactive Marketing, 15,* 13- 30.

Liu, Y. (2003, June). Developing a scale to measure the interactivity of websites. *Journal of Advertising Research,* 207-216.

Liu, Y., & Shrum, L.J. (2002). What is interactivity and is it always such a good thing? *Journal of Advertising, 31,* 53-64.

McAlexander, J.H., Schouten, J.W., & Koenig, H.F. (2002). Building brand community. *Journal of Marketing, 86,* 36-54.

McMillan, S.J., & Hwang, J. (2002). Measures of perceived interactivity: An exploration of the role of direction of communication, user control, and time in shaping perceptions of interactivity. *Journal of Advertising, 31,* 29-42.

Meuter, M.L., Ostrom, A.L., Bitner, M.J., & Roundtree, R. (2000). Self-service technologies: Understanding customer satisfaction with technology-based service encounters. *Journal of Marketing, 64,* 50-64.

Meuter, M.L., Ostrom, A.L., Bitner, M.J., & Roundtree, R. (2003). The influence of technology anxiety on consumer use and experiences with self-service technologies. *Journal of Business Research, 56,* 899-906.

Moon, Y., & Frei, F.X. (2000, May-June). Exploding the self-service myth. *Harvard Business Review,* 26-27.

Muniz, A.M., Jr., & O'Guinn, T.C. (2001). Brand community. *Journal of Consumer Research, 27,* 412-432.

Nunes, P.F., & Kambil, A. (2001, April). Personalization? No thanks. *Harvard Business Review,* 2-3.

Pine, B.J, & Gilmore, J.H. (1999). *The experience economy.* Boston, MA: Harvard Business School Press.

Pine, B.J., Victor, B., & Boynton, A.C. (1993, September-October). Making mass customization work. *Harvard Business Review,* 108-116.

Prahalad, C.K., & Ramaswamy, V. (2000, January-February). Co-opting customer competence. *Harvard Business Review*, 79-87.

Roberts, M.L. (2003). *Internet marketing: Integrating online and offline strategies*. New York: McGraw-Hill.

Sawhney, M., & Prandelli, E. (2000). Communities of creation: Managing distributed innovation in turbulent markets. *California Management Review, 42*, 24-54.

Thomke, S., & von Hippel, E. (2002, April). Customers as innovators. *Harvard Business Review*, 74-81.

Thorbjørnsen, G., Supphellen, M., Nysveen, H., & Pedersen, P.E. (2002). Building brand relationships online: A comparison of two interactive applications. *Journal of Interactive Marketing, 16*, 17-34.

Underscore Marketing. (2003). The Web offers involvement branding. Retrieved October 12, 2004, from *http://www.imediaconnection.com*

Wind, J., & Rangaswamy, A. (2001). Customerization: The next revolution in mass customization. *Journal of Interactive Marketing, 15*, 13-32.

Wyner, G.A. (2000). Customer experience on the Web. *Marketing Management*, Winter, 6-7.

Endnotes

[1] See the Advertising Research Foundation Media Model (*http://www.arfsite.org/Webpages/PrimaryPages/MediaModel_7_23_web.pdf*) for a description of the model and associated metrics for each stage that forms the basis for IAB research methodology. IAB also notes that the methodology is recognized by the European Society for Opinion and Marketing Research (ESOMAR).

[2] See Roberts (2003), especially pages 161-174 for a discussion of Internet consumer behavior and market segments.

Chapter VII

Viral Marketing:
The Use of Surprise

Adam Lindgreen,
Eindhoven University of Technology, The Netherlands

Joëlle Vanhamme,
Erasmus University Rotterdam, The Netherlands

Abstract

Viral marketing involves consumers passing along a company's marketing message to their friends, family, and colleagues. This chapter reviews viral marketing campaigns and argues that the emotion of surprise often is at work and that this mechanism resembles that of word-of-mouth marketing. Examining the literature on the emotion of surprise, the chapter next explains how a surprise is created and shared. Overall, the chapter shows how surprise can be a useful tool in a viral marketing campaign. Lastly, conclusions of interest to managers are drawn.

Introduction

Viral marketing has been described as "the process of getting customers to pass along a company's marketing message to friends, family, and colleagues" (Laudon & Traver, 2001, p. 381). Like a virus, information about the company and its products and services is spread to the customers, the customers' customers, and so on, and a huge network will rapidly be created. For example, the message of GET YOUR PRIVATE, FREE E-MAIL AT HTTP:\\WWW.HOTMAIL.COM spread to 11 million users in only 18 months (Kelly, 2000), and PayPal acquired more than 3 million users in its first 9 months of workings (De Bruyn & Lilien, 2003). Viral marketing has now gained tremendously in popularity, and world-known companies have jumped on the bandwagon: Budweiser, Kellogg's, Levi's, Nestlé, and Virgin Cinemas. However, there is still only a limited understanding of this marketing technique (Borroff, 2000). Consider also Helm (2000) who contends that "definitions and backgrounds are hardly focused" (p. 158). Other authors argue in a similar vein (Brodin, 2000; De Bruyn & Lilien, 2003; Diorio, 2001). For example, De Bruyn and Lilien (2003) posit that "it is difficult to...explain why and how [viral marketing] works" (p. 4).

Brewer (2001) agrees that viral marketing – otherwise known as referral marketing (Brewer, 2001; Murphy, 2002) - is "the process of getting customers to pass along a company's marketing message to friends, family, and colleagues" (Laudon & Traver, 2001, p. 381), but adds that the cost of viral marketing is only incremental. Viral marketers make use of consumer-to-consumer communications, which are both rapid and cheap (Laudon & Traver, 2001). Helm (2000) understands viral marketing as a type of advertising that is almost like an online version of word-of-mouth advertising (c.f., Beckmann & Bell, 2000). Viral marketing does not necessarily have to take place over the web. However, with the Internet facilitating consumer interconnections dramatically (De Bruyn & Lilien, 2003) viral marketing has become a buzzword for viral marketing campaigns happening through the Internet. Viral marketing can be used for both promoting and distributing products (Helm, 2000).

Brodie (2001) argues that one of the oldest examples of a viral marketing campaign is the Gospel and the Christian missionaries preaching the life and teaching of Jesus. One of the probably most famous viral marketing campaigns is that of *The Blair Witch Project* where the budget for the movie's release was just $2.5 million, but the movie grossed $245 million in worldwide box office sales. Before the movie was released, Artisan Entertainment, the maker of the movie, created much interest in the movie by giving it the air of a documentary and supporting it with an Internet site. People then talked about *The Blair Witch Project* and referred their friends to the site (Bernard & Jallat, 2001; Rasmusson, 2000). Other famous viral marketing campaigns include Anheuser-Busch's Budweiser "Wassup," Lees Buddy Lee mascot, Kinetix's free Dancing Baby animation sample, Blue Mountain Arts' greeting card service, and Virgin Mobile's Red Academy campaign. Interestingly, Seth Godin's book *Unleash the Ideavirus*, itself about viral marketing, was originally made available free-of-charge on http://www.ideavirus.com, and readers could send a full electronic version of the book to their friends.

So What is Viral Marketing?

Through viral marketing, products are diffused fast because users are taken to the home page by the person who referred them to the home page (Helm, 2000). Viral marketing involves (1) users who know each other (compare the Refer-a-Friend program of *Half.com*; Reichheld & Schefter, 2000) and (2) users who do not know each other (compare the sites of *Epinions.com* and *ConsumerReports.org* that provide objective product reviews to interested consumers; Laudon & Traver, 2001). In good viral marketing campaigns, the product is only on offer on the Internet and contains real value for the potential customer. Also, a company chooses very carefully which people should first pass on the viral marketing message, as the creation of viral networks rests on these people (Bannan, 2000; Harvard Management Update, 2000; Helm, 2000).

Viral marketing strategies can be classified according to the degree of requiring the customer's activity in forwarding the viral marketing message (Helm, 2000): low and high integration strategies. The former strategies include "Send this story to a friend" icons and can be used for Web-hosted address books, calendars, list servers, newsgroup readers, and greeting card services. The latter strategies require the active participation of the customer in reaching new users who may have to download special programs (Helm, 2000; Jurvetson & Draper, 1998). De Bruyn and Lilien (2003) echo this when they discuss intentional viral message dissemination (e.g., PayPal or the "Recommend It" that is known from numerous Web sites) and unintentional viral message dissemination (e.g., Hotmail.com).

Viral marketing-based online referrals are usually sent to people who are not actively seeking information and therefore "not a priori willing to pay attention to them" (De Bruyn & Lilien, 2003, p. 4). Still, studies suggest that viral marketing is effective as a means of drawing high response rates. It is responsible for almost 60% of customer visits to Web sites (Schweinsberg, 2000). Up to 15% of shoppers who receive a viral message follow the links with the highest rate in business-to-business markets (Rigby, Siddle & Chu, 2000). This is reflected in the value of companies that build their success on viral marketing. ICQ was sold to AOL.com for $287 million in 1998; Microsoft acquired Hotmail.com for $400 million; and Yahoo bought GeoCities! for $4 billion in stock (De Bruyn & Lilien, 2003; Helm, 2000).

E-commerce was originally fought on price, which meant that the middleman was cut out so that margins could be stripped. When the market became flooded with numerous cut-price competitors, e-commerce examined ways of retaining customers. Clicks-and-mortar companies should now, Rigby, Siddle, and Chu (2000) posit, seek to win hearts meaning that companies "must learn to delight their customers through excellent service, online and off" (p. 1). Interestingly, from Table 1, which summarizes what has been identified as driving viral marketing campaigns, it is seen that the ability to elicit affective reactions and/or the personal affective attachment/involvement seems to underline most of these programs. This chapter thus argues that emotions are key drivers of viral marketing campaigns and that surprise is a particularly important emotion to consider if a campaign is to be successful. This is in line with Rigby, Siddle, and Chu (2000), as well as Rust and his colleagues, namely that companies "need to move beyond mere satisfaction to customer delight" (Rust, Zahorik & Keiningham, 1996, p. 229) and that the features that have "the capacity to delight are those that are…surprisingly pleasant" (Rust & Oliver, 2000, p. 87). Table 1 gives an overview of the different mechanisms that have been suggested in the literature. Some companies believe that viral marketing works for all kinds of products (Brown, 2001); other companies think that viral marketing is appropriate as a nice tool, whereas its value as a mass marketing tool should be assessed on a project-by-project basis (Regan, 2002).

Table 1. Mechanisms behind viral marketing

Mechanism(s)	Source and Explanation	Mechanism(s)	Source and Explanation
• Entertainment, amusement, irritation	*Splash of Paint*: People are directed to the company's Internet site by entertaining, amusing, and/or irritating them.	• Coolness, fun • Second-to-none offer	*Virgin Atlantic*: Customers pass on the message when they think it is cool or fun, or if the offer is second to none.
• Fun, quirk, amusement • Specific and relevant to the person	*Claritas*: Viral marketing campaigns should be funny, quirky, or amusing, or something that is very specific and relevant to the individual customer.	• Violence, pornography, irreverent humor	*Clark McKay and Walpole Interactive*: The messages drawing highest response rates are those who have elements of violence, pornography, or irreverent humor.
• Fun, humor, excitement (jokes, games)	*Fabulous Bakin' Boys*: Its Web site supports the muffin products with flash animation sites, fun, jokes, as well as games that people can download and forward to their friends.	• Comic strips, video clips	Comic strips and video clips grab the attention of people who then forward the content to their friends (Harvard Management Update, 2000).
• Emotional elements	Internet strategies must have high levels of emotional content, including interactivity, the ability to involve other people, chat rooms, and the creation of online community (Barnes & Cumby, 2002).	• Contests and humor • Important advice	Contests and humor are important elements in successful campaigns, which can also be successful if they have important advice to customers (Zimmerman, 2001).
• Nature of the industry • Online tenure of the audience • Topic	*Sage Marketing & Consulting Inc.*: The success of viral marketing is dependent on (1) the nature of the industry that the company is in; (2) the online tenure of the audience; and (3) the topic. People are more likely to pass on information about products like entertainment, music, Internet, and software.	• Controversy	A company gains publicity when the media writes about controversy on its Web sites, and competitors will have to deal with the company. But such word-of-mouth marketing can be dangerous because dissatisfied customers are more likely to share their negative experience than satisfied customers (Wilson & Abel, 2002).
• False, deliberately deceptive information • Popularly believed narrative, typically false • Anecdotal claims • Junk	So-called "urban legends and folklore" can be organized as (1) false, deliberately deceptive information; (2) popularly believed narrative, typically false; (3) anecdotal claims, which may be true, false, or in between; (4) and junk. Such stories are frequently forwarded to friends, family, and colleagues (Urban Legends, 2002).	• Fun, intrigue, value • Offer of financial incentives • Need to create network externalities	People pass on messages if they find the product benefits to be fun, intriguing, or valuable for others; if they are given financial incentives for doing so; or if they feel a need to create network externalities (De Bruyn & Lilien, 2003).

The Emotion of Suprise

Our understanding of the emotion of surprise in a marketing context is rather limited and originates largely from the psychology literature (compare Lindgreen & Vanhamme, 2003). Indeed, only few researchers have attempted to examine the effects of surprise on relevant different marketing variables, such as word-of-mouth, customer satisfaction, and customer retention (e.g., Derbaix & Vanhamme, 2003; Vanhamme & Snelders, 2001; Vanhamme, Lindgreen & Brodie, 1999).

Surprise is a short-lived emotion characterized by a specific pattern of reactions (Reisenzein, Meyer & Schützwohl, 1996). (For a detailed review, see Vanhamme & Snelders, 2001.) When products, services, or attributes are unexpected or misexpected, they elicit surprise (Ekman & Friesen, 1975). *Unexpected* denotes vague and not well-defined expectations about a product, service, or attribute while *misexpected* denotes precise expectations about a product, service, or attribute that do not occur, however. Both unexpected and misexpected products, services, or attributes form a schema discrepancy that is a type of private, normally informal, inarticulate, and unreflective theory about the nature of objects, events, or situations (Schützwohl, 1998). The appropriateness of a personal schema is continuously (and relatively unconsciously) checked against the surrounding environment (Schützwohl, 1998). Surprise is elicited as soon as reality diverges from the schema and results in processes that aim at eradicating the schema discrepancy, which can lead, if necessary, to updating of the schema (Reisenzein, Meyer & Schützwohl, 1996).

Despite the emotion of surprise being neutral, it is frequently succeeded by another emotion that colors it positively (e.g., surprise + joy) or negatively (e.g., surprise + anger), which explains why people talk about good/positive surprises and bad/negative surprises (Ekman & Friesen, 1975). Here the terms *positive surprise* and *negative surprise* also refer to the blend of surprise with a subsequent positive and negative emotion, respectively. It is also worth mentioning that surprise, through its intrinsic arousal, amplifies subsequent affective reactions (Desai, 1939), implying that people who feel joy (anger) after having been surprised will feel more joyful (be angrier) than if they had not previously been surprised.

"Emotionize" Viral Marketing

The referral mechanism behind viral marketing resembles that of word-of-mouth marketing (e.g., Buttle, 1997; Helm, 2000; Lindberg-Repo & Grönroos, 1999): "positive WOM [word-of-mouth] occurs when good news testimonials and endorsements desired by the company are uttered [and] negative WOM is the mirror image" (Buttle, 1997, p. 4). But Bansal and Voyer (2000) posit that "there is surprisingly little empirical research that examines [word-of-mouth] 'procedural' aspects" (p. 166). The chapter now proceeds to examine, first, the similarities and differences between viral marketing and word-of-mouth and, second, the potential benefits of employing the emotion of surprise as a viral marketing driver. Beckmann and Bell (2000) found differences between viral marketing and word-of-mouth marketing that can have implications for the applications of each approaches. For example, viral messages are spread more quickly to a larger number of people than word-of-mouth messages. Viral messages largely rely on visual stimuli, both text and imagery, whereas word-of-mouth messages most often are verbal or face-to-face communication. Also, it is easier for a company to control the nature and content of a viral message because world-of-mouth messages depend on the sender, meaning that the message can become compromised over time. Lastly, word-of-mouth marketing typically involves two-way communication with receiver feedback, and the likelihood that the receiver will attend to the whole message is greater in word-of-mouth marketing.

That viral marketing draws high response rates is in agreement with findings on how word-of-mouth influences consumers' choices and purchase decisions, shapes their expectations, pre-usage attitudes, and post-usage perceptions of products and services, as well as the fact that the influence of word-of-mouth often is greater than more impersonal communication, such as advertising. For a review, we refer to Lilien and his colleagues (De Bruyn & Lilien, 2003) who have also reviewed research on how word-of-mouth works: Consumers engage in word-of-mouth when they are highly satisfied or dissatisfied, when they feel committed to a company, or when a product or service is novel. Further, word-of-mouth is more likely to be at play if consumers know little about a product category, or if they are deeply involved in a purchase decision. Lastly, the influence of personal sources of information is higher than that of other sources because of source expertise, tie strength, demographic similarity, and perceptual affinity.

According to Westbrook (1987), salient positive or negative emotions stimu-late word-of-mouth communication, and the influence of positive and negative emotions is not mediated by satisfaction or its cognitive antecedents (i.e., expectation and disconfirmation beliefs). In the same vein, Maute and Dubé (1999) use a mental simulation of a core service failure and show that emotional responses accounted for a large part (about 30%) of the explained variance of word-of-mouth intentions. Derbaix and Vanhamme (2003) also present evi-dence that the emotion of surprise has a strong influence on word-of-mouth (that is not fully mediated by subsequent positive or negative emotions). Since the differences observed between word-of-mouth marketing and viral market-ing (Beckmann & Bell, 2000) do not relate to the emotional content or the effect of the message, we argue that emotions–and more specifically, surprise–are likely to play an important role in creating an effective viral marketing campaign.

The influence of positive and negative emotions and the emotion of surprise on viral marketing messages can be related to the phenomenon of "social sharing of emotions." This is defined as "a phenomenon involving (1) the evocation of the emotion in a socially shared language and (2) at least at the symbolic level at some addressee" (Rimé et al., 1992, p. 228) (e.g., to another person) in a diary. People, who experience everyday life emotions, initiate communication processes during which they share parts of their private experiences with social partners (i.e., family, friends, and colleagues). Only about 10% of emotional experiences are kept secret and never socially shared with anyone (Rimé et al., 1992). There is also evidence that the more disruptive an event is, the sooner and the more frequently it is shared. Social sharing of emotions is also positively related to the intensity of the emotions (Rimé et al., 1998). Further, the process does not stop after the first social sharing. Instead, people who are part of the social sharing of emotions initiate a secondary social sharing process; the more intense the emotional event has been, the more people engage in this secondary social sharing process (Christophe & Rimé, 1997), and an entire network is created.

Surprise constitutes a spectrum of changes, such as interruption of ongoing activities, focus of attention, and physiological changes, all of which are disruptive. Surprise also elicits substantial cognitive burden (causal search, causal attribution, schema updating, etc.) that can lead to additional interactions with other people to the extent that these interactions offer help to the individual in alleviating this burden (Söderlund, 1998). Surprise can thus be considered as a disruptive emotion that has a high likelihood of inducing a social sharing process like viral marketing.

There might be at least one additional reason (other than sharing one's emotions) in word-of-mouth activity elicited by surprising experiences (Derbaix & Vanhamme, 2003): surprised speakers could engage in such activity on the basis of the perceived utility to their audience. Most people assume that what is surprising for themselves will also be new and thus useful information to other people (e.g., the utilitarian side of the word-of-mouth activity to others.) In viral marketing, it means that surprised consumers create larger networks than nonsurprised consumers if they believe that the message will be of some utility to the receivers (e.g., the receivers will experience surprise and/or positive emotions or gain something from the message). That is, the positive influence of surprise on the size of the viral network is moderated by the utility to other people receiving the message.

Research Findings

A questionnaire was distributed to students following a business master program. All 45 students gave back the questionnaire; they were not allowed to discuss with each other while filling it out. The introduction text to the first part of the questionnaire was "Very often, we receive e-mails from, or about, companies or organizations, or their products and/or services, e-mails that we forward to other family, friends, and colleagues. Please write down an e-mail that you forwarded through e-mail to a large number of family, friends, and colleagues if you had received it through e-mail." The introduction text to the second part of the questionnaire was "Now please write down an e-mail–received from, or about, a company or organization, or its products and/or services–that you did not forward through e-mail, or that you forwarded only to a very limited number of people." Two independent judges first organized the findings and then coded what seemed to be success factors in viral marketing.

We found that students generally forward e-mails either if these involve something for free (e.g., a free ticket or product such as a game) or at a reduced price (e.g., a trip for a low cost), or if the e-mails contain jokes or games. An example of a free game is an online version of the famous con trick Three Card Monte where the audience is asked to find the card, which is the red queen. In real life, the audience never can find the red queen and thus loses the bet. Since the trick is based on slight of hand (Lindgreen, 1995), the viewers are surprised that the online version (i.e., a computer-mediated setting) knows which card the audience will pick and therefore forward the free game to people whom they know will appreciate it! Other forwarded e-mails are about a virus alert or health problems caused by certain products (or any similar serious thing). Students also forward an e-mail when it has to be sent to a number of people in order not to have any bad luck (or in order to bring these people good luck). (Other students, though, wrote that they do not forward this kind of e-mail because they are not superstitious.) They also forward e-mails about a missing child, a charity event, or other similar things (e.g., collection of clothes and food). One case in point is e-mails on how women were treated in Afghanistan at the time of the Taliban government. In contrast, students tend not to forward e-mails when these are pure advertising where there is nothing to be gained, racist messages, or untrustworthy messages. This includes the obviously deceitful messages from certain African countries where the sender is asking for help in transferring a huge amount of money (typically in the order of $10 million).

These preliminary findings thus corroborate the suggestion that a large part of forwarded e-mails have elicited (positive, but also negative) surprise (e.g., a free present, a new game, or a price reduction) and were considered to be of some utility to the receivers. Viral marketing is based on word-of-mouth that takes place between people, and a sender of a viral marketing e-mail must thus see the e-mail as benefiting the receiver. Friends know what surprises their friends, so they are the best people to forward e-mails that will be successful. If a company can reach some customers who get surprised, it stands a good chance that the surprising e-mail will be forwarded to more customers. People forward e-mails if they believe that the receivers will benefit from it (e.g., the free game), or if they think that the society at large will be better off (e.g., the Afghanistan case). This is why pure advertising e-mails are not forwarded: they benefit only the sender; e-mails advertising a product or service must have something else attached in order to be forwarded. Further, the findings suggest that trust (and, to some extent, superstition) could act as an additional moderator of the surprise-viral marketing relationship. In summary, the more surprised customers are, the more likely they are to generate a large viral network (a) if they believe it will benefit the receiver and (b) if they trust the source of the message.

Managerial Practice

Some guidelines exist for launching a viral marketing campaign, and Figure 1 summarizes the steps that should be undertaken by marketers. We will discuss some of these guidelines.

Figure 1. Steps in order to create a viral marketing campaign

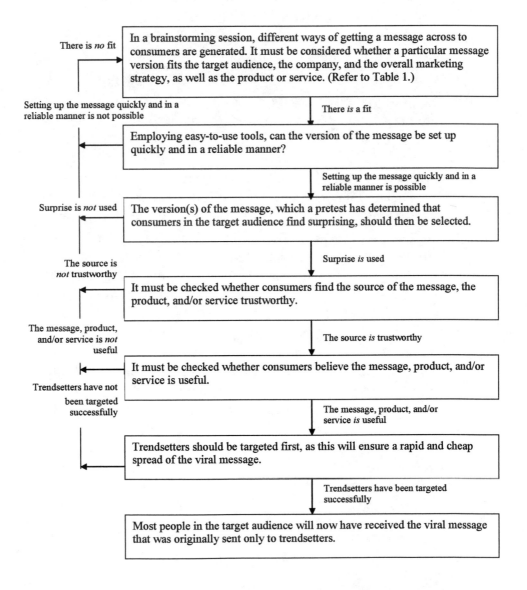

Companies should understand the target audience and consider whether a message fits it, the company, and the marketing strategy; use contests and promotions; test an intended viral marketing campaign before implementing it fully; integrate the campaign into the overall marketing strategy; provide customers with information that enables them to understand the company's value proposition; build in exit barriers; and measure the rates of customer advocacy. Companies should avoid spamming and the need to come up with incentives in the long run. Viral marketers often benefit if they first target the influential people within their target audience. Such trendsetters are said to account for 8% of Internet users, and they influence about eight other people. Further, the tools which a campaign can employ (including e-mail, chat rooms, bulletin boards, gifts, wish list registers, or Web pages) must be quick, reliable, and easy to use (Thevenot & Watier, 2001). This means that companies should avoid graphics that take a long time to download (quick issue) or forms that take a long time to fill out (easy-to-use issue), as well as programs that are prone to bugs (reliable issue). Some products or services have to be accepted by the initial consumer and his friends, family, and colleagues before being of value (e.g., ICQ that allows people to communicate with each other through the Internet only if they use this particular computer program). People generally engage in viral marketing when they:

- receive freebies (e.g., software programs, services);
- are given (monetary) incentives including discounts, coupons, tickets, and contests;
- are concerned about an issue, such as the environment, the war in Iraq, or debt relief to poor countries; and
- need to create network externalities.

A viral marketing campaign's success also depends on whether or not the message is:

- funny, humorous, amusing, or cool;
- quirky, intriguing, or even irritating and controversial – or, in extreme cases, violent or pornographic; and
- offers important or second-to-none advice that is both specific and relevant to the person.

Products should be services built around the viral message, not the other way around (Godin, 2001). Obviously, some elements are more suitable for a business-to-consumer setting, while other elements work well in a business-to-business environment. Apparently, some industries, including entertainment, music, software, and the Internet, are especially suited to viral marketing campaigns (Thevenot & Watier, 2001). This is interesting because e-commerce has been most successful for books, music/CDs, electronics, holidays/travels, and PC hardware and software (Taylor Nelson Sofres Interactive, 2002). Lastly, the use of emotions is key in constructing a campaign that works (compare Plutchik, 1980). Some companies are successful, while others fare badly and charge a fee for their products or service which they should only do if people can identify the benefits or if a large user base has been established. Also, the fee must be for add-ons to the product or service, not for the basic one (Thevenot & Watier, 2001).

Conclusions

This chapter identified different drivers of viral marketing and showed that they often contain surprising elements. Next, the chapter examined the literature on the emotion of surprise and explained how surprise is created and how people share surprising experiences. In making a parallel to the word-of-mouth literature, it was illustrated how the emotion of surprise can be a useful tool in a viral marketing campaign. Based on the literature review and exploratory research, it was suggested that viral marketing campaigns be based on surprising messages that benefit the receivers' friends, family, and relatives in some way (e.g., create surprise, be pleasant, etc.) in addition to being perceived as originating from a trustworthy source. Lastly, a step-by-step design for creating a viral marketing campaign using surprise was presented. There is an intimate relationship between the original viral receivers and the subsequent ones, and this means that a company that succeeds in reaching some consumers from its target audience *and* surprising them will stand a good chance of having created a successful viral network.

References

Bannan, K.J. (2000, June). It's catching. *IQ Interactive News, 5*, 20-27.

Bansal, H.S., & Voyer, P.A. (2000). Word-of-mouth processes within a services purchase decision context. *Journal of Service Research, 3*(2), 166-177.

Barnes, J.G., & Cumby, J.A. (2002). Establishing customer relationships on the Internet requires more than technology. *Australasian Marketing Journal, 10*(1), 36-46.

Beckmann, S.C., & Bell, S. (2000). Viral marketing = a word-of-mouth marketing on the Internet? (And a case story from the Nordic countries). In A. O'Cass (Ed.), *Visionary marketing for 21st century problems*. Gold Coast, Queensland, Australia: Griffith University.

Bernard, G., & Jallat, F. (2001, March). Blair Witch, Hotmail et le marketing viral. *L'Expansion Management Review*, 81-92.

Borroff, R. (2000, November 20). Viral marketing. *Precision Marketing*.

Brewer, B. (2001). Refer a friend: Best practices for referral marketing with e-mail. *The DMA Insider,* Summer, 30-34.

Brodie, R. (2001). Viral marketing. *Meme Update*, No. 33. Retrieved October 12, 2004, from *http://www.memecentral.com/mu/mu0033.htm*

Brodin, O. (2000, September-December). Les communautés virtuelles: Un potentiel marketing encore peu exploré. *Décisions Marketing, 21*, 47-56.

Brown, C. (2001, June 29). Big brands catch the viral marketing bug. *Precision Marketing*.

Buttle, F.A. (1997). I heard it through the grapevine: Issues in referral marketing. In M. Christopher & A. Payne (eds.), *Proceedings of the 5th International Colloquium in Relationship Marketing*. Cranfield, England: Cranfield School of Management.

Christophe, V., & Rimé, B. (1997). Exposure to the social sharing of emotion: Emotional impact, listener responses and secondary social sharing. *European Journal of Social Psychology, 27*, 37-54.

De Bruyn, A., & Lilien, G.L. (2003). *Harnessing the power of viral marketing: A multi-stage model of word of mouth through electronic referrals* (Working Paper). Pennsylvania State University, Department of Business Administration. Retrieved October 12, 2004, from *http://www.arnaud.debruyn.info/research/papers/viralmarketing.pdf*

Derbaix, C., & Vanhamme, J. (2003). Inducing word-of-mouth by eliciting surprise: A pilot investigation. *Journal of Economic Psychology, 24*(1), 99-107.

Desai, M.M. (1939). Surprise: A historical and experimental study. *British Journal of Psychology: Monogr*, Suppl. 22.

Diorio, S. (2001, February 16). How to catch on to viral marketing. *ClickZ Today*. Retrieved October 12, 2004, from *http://www.clickz.com/mkt/onl_mkt_strat/article.php/837321*

Ekman, P., & Friesen, W. (1975). *Unmasking the face*. Englewood Cliffs, NJ: Prentice Hall.

Godin, S. (2001). *Unleashing the ideavirus*. New York: Hyperion.

Harvard Management Update (2000). Keeping up: Viral marketing. *Harvard Management Update, 5*(6), 9.

Helm, S. (2000). Viral marketing – Establishing customer relationships by word-of-mouse. *Electronic Commerce and Marketing, 10*(3), 158-161.

Jurvetson, S., & Draper, T. (1998). Viral marketing. *Business 2.0, 1*.

Kelly, E. (2000, November 27). This is one virus you want to spread. *Fortune*, 297-300.

Laudon, K.C., & Traver, C.G. (2001). *E-commerce: Business, technology, society*. Boston: Addison-Wesley.

Lindberg-Repo, K., & Grönroos, C. (1999). Word-of-mouth referrals in the domain of relationship marketing. *Australasian Marketing Journal, 7*(1), 109-117.

Lindgreen, D. (1995). *Tre-kort spillet*. Horsens, Denmark: Pegani.

Lindgreen, A., & Vanhamme, J. (2003). To surprise or not to surprise your customers: The use of surprise as a marketing tool. *Journal of Customer Behaviour, 2*(2), 219-242.

Maute, M.F., & Dubé, L. (1999). Patterns of emotional responses and behavioral consequences of dissatisfaction. *Applied Psychology, 48*(3), 349-366.

Murphy, D. (2002, January 18-20). The friendly virus. *Sales Promotion*, 18-20.

Plutchik, R. (1980). *Emotion: A psychoevolutionary synthesis*. New York: Harper & Row.

Rasmusson, E. (2000, June 18). Viral marketing: Healthier than it sounds. *Sales & Marketing Management, 18*.

Regan, J. (2002, February 1). Just stop and think before you start your next viral campaign. *Precision Marketing*.

Reichheld, F.F., & Schefter, P. (2000). E-loyalty: Your secret weapon on the Web. *Harvard Business Review, 78*(4), 105-113.

Reisenzein, R., Meyer, W-U., & Schützwohl, A. (1996). Reactions to surprising events: A paradigm for emotion research. In N.H. Frijda (ed.), *Proceedings of the 9th Conference of the International Society for Research on Emotions* (pp. 292-296), Toronto, Canada: ISRE.

Rigby, D., Siddle, R., & Chu, J. (2000). *eStrategy brief: Delighting customers online*. Chicago: Bain & Company and Cambridge, Massachusetts: Mainspring.

Rimé, B., Finkenhauer, C., Luminet, O., Zech, E., & Philippot, P. (1998). Social sharing of emotions: New evidence and new questions. In W. Stroebe & M. Hewstone (eds.), *European review of social psychology, 9* (pp. 145-189). Chichester, England: John Wiley & Sons.

Rimé, B., Philippot, P., Boca, S., & Mesquita, B. (1992). Long lasting cognitive and social consequences of emotion: Social sharing and rumination. In W. Stroebe & M. Hewstone (Eds.), *European review of social psychology, 3* (pp. 225-258). Chichester, England: John Wiley & Sons.

Rust, R.T., & Oliver, R.L. (2000). Should we delight the customer? *Journal of the Academy of Marketing Science, 28*(1), 86-94.

Rust, R.T., Zahorik, A., & Keiningham, T.L. (1996). *Service marketing*. New York: HarperCollins.

Schützwohl, A. (1998). Surprise and schema strength. *Journal of Experimental Psychology, Learning, Memory and Cognition, 24*(5), 1182-1199.

Schweinsberg, K. (2000, May 10). Viral marketing 101 – Getting started. *Digitrends.net*. Retrieved October 12, 2004, from *http://www.digitrends.net/marketing/13640_10856.html*

Söderlund, M. (1998). Customer satisfaction and its consequences on customer behaviour revisited: The impact of different levels of satisfaction on word-of-mouth, feedback to supplier and loyalty. *International Journal of Service Industry Management, 9*(2), 169-188.

Taylor Nelson Sofres Interactive (2003, October). *Global eCommerce report.* Retrieved October 12, 2004, from *http://www.tnsnipo.nl/onderzoek/gratis/persvannipo/pdf/rapport_ger2002.pdf*

Thevenot, C., & Watier, K. (2001, May). Viral marketing. *Sovereign Music.* Retrieved October 12, 2004, from *http://www.sovereignmusic.com/viralmarketing.html*

Urban Legends. (2002). Retrieved October 12, 2004, from *http://urbanlegends.about.com*

Vanhamme, J., Lindgreen, A., & Brodie, R.J. (1999). Taking relationship marketing for a joyride: The emotion of surprise as a competitive marketing tool. In J. Cadeaux & M. Uncles (Eds.), *Marketing in the third millennium.* Sydney, New South Wales, Australia: University of New South Wales.

Vanhamme, J., & Snelders, D. (2001). The role of surprise in satisfaction judgments? *Journal of Consumer Satisfaction, Dissatisfaction and Complaining Behavior, 14,* 27-45.

Westbrook, R.A. (1987). Product/consumption-based affective responses and postpurchase process. *Journal of Marketing Research, 24*(3), 258-270.

Wilson, S.G., & Abel, I. (2002). So you want to get involved in e-commerce. *Industrial Marketing Management, 31*(4), 85-94.

Zimmerman, E. (2001, February). Catch the bug. *Sales & Marketing Management,* 78-82.

Chapter VIII

Retailer Use of Permission-Based Mobile Advertising

Jari Salo, University of Oulu, Finland

Jaana Tähtinen, University of Oulu, Finland

Abstract

This chapter shows how retailers – without any previous experience or education – utilize a new mobile advertising channel. The chapter focuses on permission-based mobile advertising and the specific features that should be considered when designing and targeting mobile advertising. The empirical part of the chapter analyzes data from a field trial where Finnish retailers were able to use mobile advertising. The empirical data is obtained through the use of content analysis. Data analysis explores whether the retailers positioned the m-adverts to target either individuals or groups and whether the content of the m-adverts reflected and utilized the medium's specific features. The results of this analysis suggest that both retailers and advertising agencies have to learn how to use m-advertising as a new media. Conclusions of the chapter suggest ways to fully utilize the potential of mobile advertising.

Introduction

M-advertising, or wireless advertising, has two different meanings in marketing literature. First, the term refers to advertisements that move from place to place. Buses, trucks, trains, trams, and taxis provide ideal settings for this type of m-advertising (Goldsborough, 1995; Hume, 1988). Second, m-advertising refers to adverts sent to and received on mobile devices (i.e., cellular phones, Personal Digital Assistants (PDAs), and other handheld devices that people carry with them). The context specificity allows advertisers to send targeted and personalized m-adverts to consumers on the move, hence an alternative term, location based commerce (Turban et al., 2002).[1] In this chapter, m-advertising refers to the second meaning of the term.

M-advertising can be seen as a part of m-commerce (Clarke & Flaherty, 2003; Mennecke & Strader, 2003). M-commerce is referred to as a radically different direction away from traditional commerce (Choi, Stahl & Whinston, 1997), as it offers a way in which to sell and distribute the retailers' digital products or services to customers through a mobile device. Thus, it can be argued that m-advertising is also radically different from traditional ways of advertising. M-advertising enables not only the sending of unique, personalized, and customized adverts (Turban et al., 2002) but also the ability to engage consumers in interaction with the sender of the message.

In the future, any retailer will be able to make use of m-advertising channel. With m-advertising, consumers can be reached quickly at anytime, although the commercial transactions would be limited to store opening hours. Thus, it is no surprise that m-advertising is predicted to become the second largest form of m-commerce in the year 2005, reaching more than $6 million in Europe alone (Durlacher, 2000).

Although the importance of m-advertising is clear, research into the area is scarce. Empirical research of context aware m-advertising in a real life setting presents quite a challenge. First of all, the technology has to be readily available, and second, the consumers have to be familiar with mobile devices. In the city of Oulu, Finland, the above mentioned prerequisites are in place. The SmartRotuaari research project offers the brick-and-mortar retailers in downtown Oulu the infrastructure and software solutions they need to use context aware m-advertising. In Finland, consumers' mobile phone subscriptions reached 84% in 2002 (Ministry of Transport and Communications Finland, 2003). In addition, more than 30% of users younger than 35 years have received m-advertising in the form of an SMS, and over 20% of all users have made mobile payments (*www.opas.net/suora/*).

Although the technology is available and used, Finnish legislation restricts m-advertising to some extent, as it allows only permission-based m-advertising. Without consumers' permission, only information that is related to the administration of the customer relationship (e.g., the customer's flight to Dallas is late) can be sent via mobile devices. Therefore, this empirical research is limited to permission-based m-advertising.

Technology and the consumers are not the only variables needed in order to effectively use m-advertising. The advertisers and their advertising agencies have to know how to design m-adverts as well as how to use this new advertising channel. The objective of this chapter is to analyze how Finnish retailers use permission-based context aware m-advertising when offered a chance to try the new channel.

Based on existing research as well as on the empirical data, the chapter continues with a discussion on the unique features of m-advertising. The data shows that most of the m-adverts did not utilize the features that make m-advertising unique and effective. The chapter concludes by suggesting ways that retailers could improve their use of m-advertising in order to fully harness the power that it offers.

Factors Influencing the Success of M-Advertising

Based on existing research and the empirical data gathered for this study, we suggest a framework that describes the factors that influence the success of retailers' use of m-advertising. As pictured in Figure 1, the influencing factors are related to m-advertising as a media and to the receiver of the messages (i.e., the individual customer's goals in using the mobile device). The above mentioned factors set demands for retailers' use of the m-advertising. M-advertising should be used to deliver adverts which differ from traditional ones. Traditional advertising is designed for and delivered to target groups, but m-advertising should be designed for – as it is delivered to – target individuals. Unless the target individual perceives permission-based m-advertising positively, that individual will deny the company and any other company attempting to advertise the permission that they require. Therefore it is vital for a m-advertiser to meet the requirements that are set for the targeting and the content of the m-adverts.

Personal Nature of M-Advertising

M-advertising is on a par with personal selling. Mobile devices, especially mobile phones, are highly personal devices with personally selected or even self-composed ringing tones, individually tailored covers or general appear-

Figure 1. The factors influencing the success of m-advertising

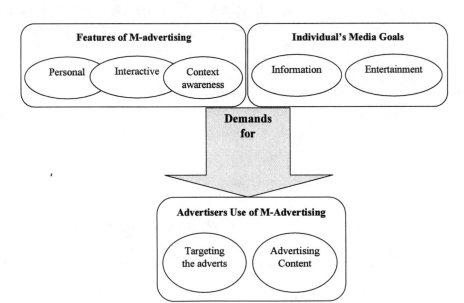

ance, and additional decorations, not to mention the content of the phone which often includes information about friends as well as a personal calendar. Moreover, the users carry their device almost everywhere and at all times. Thus, the personal nature of the device is transferred to the information that is sent and received through the device. Therefore, m-advertising is not for masses but for individuals.

Interactive Nature of M-Advertising

The mobile device allows m-advertising to be highly interactive (i.e., the parties can interact with each other, the communication medium, and the messages) (Liu & Shrum, 2002). The customer may reply to the advert by phoning, sending an SMS, MMS, or an e-mail, or logging into the advertiser's Web page using the mobile device. In addition, the customer may forward the ad to friends. This type of viral marketing, of course, is very beneficial for the advertiser, as when a customer forwards the m-advert, they then become the senders of the message and therefore the message gains credibility. However, a customer will only forward m-adverts considered to be of some value, either monetary or entertainment. Interaction also has a downside, as customers may easily forward negative information about the advertiser.

Context Awareness of M-Advertising

The first context to be taken into account is the device to which the m-advert is distributed. Unless the message is tailored to the terminal, the receiver will encounter problems in receiving and understanding the message. For example, m-adverts containing video and music clips cannot be sent to mobile phones which do not support them. Different mobile phone models and even the same models may show an advert differently on the screen, depending on the phone's software version. Even if such problems are avoided, the devices have a relatively small screen size, limited screen colors, and battery time.

In addition, the technology used in building m-advertising systems enables m-advertising to be aware of the context of the receiver. M-advertising can be related to location, weather, and time, all of which can be used in targeting adverts. For example, by using location awareness, the m-advertiser is able send an advert if the customer walks by the retailer's shop. Moreover, since people carry their devices with them almost all the time, m-advertising is an extremely rapid media. The adverts are delivered, read, and acted on almost immediately. Retailers can thus receive the first feedback from their campaigns while they are still running.

Individuals' Media Goals

Individuals' goals are often referred to as a person's cognition of what is pursued in a particular situation and associated with an inner state of arousal (compare Eysenck, 1982; Pervin, 1989). Thus, an individual's media goal is the cognition of the processing goal the individual is pursuing when attending to the mobile device (compare Juntunen, 2001), which in this case is the medium for m-adverts. The type of goal receivers are trying to achieve by using a mobile device also influences their processing of the advert. If the users' media goals are information, the customer will be more interested in adverts that provide them with relevant information. On the other hand, if the customers' goals are directed more towards entertainment, which is very probable, at least among the younger consumers, they will enjoy adverts that are entertaining and provide experiential satisfaction through aesthetic pleasure, emotional stimulation, or social experience (compare Barwise and Strong, 2002). However, a consumer may wish to achieve both kinds of goals at the same time, and the relative importance of the types may change according to the individual's situation at the time.

We have now discussed the features that make m-advertising a unique form of advertising, namely, its personal and interactive nature and the context aware-ness. Moreover, the way the users' are using their mobile device also influences the way they perceive the adverts they receive. Since m-advertising is so personal, it sets new kinds of demands when planning advertising. In the following section, we will move into considering how to plan m-advertising. There are two main decisions to be made when planning any advertising campaign: to whom do we want to talk (i.e., targeting the advert), and what do we want to say to them (i.e., the advertising content).

Targeting the Adverts

It is possible to target m-advertising if the retailer can make use of the user-specific information, which is added to the m-advertising service system. This can be done through two different but complementary ways. First, each user, when granting the permission to send adverts also fills in a user profile. That may include demographic details, user's current mood (e.g., is the user hungry, looking for fun, shopping, etc.) and areas of personal interest (e.g., fashion, movies, food, hunting, reading, etc.). Users also input presence status that tells if the consumer is idle or willing to receive adverts. All this can be done directly from the user's mobile device. Second, the retailer may use existing data from the company's Customer Relationship Management (CRM) database.

Moreover, the content awareness allows retailers to target adverts according to the local weather conditions. The system may obtain up-to-date weather information from a local weather station via Internet. Thus, m-advertising allows retailers to take advantage of changing weather conditions by sending a sunglass advert only when the sun is shining.

In addition, time can be used in targeting. Depending on the time of the day, retailers can send m-adverts that address the needs of customers at a particular point in time. In the morning, restaurants can send special breakfast offers, and in the evening, they can send discount coupons for a dinner if there are seats available. For pub and restaurant owners, the time range used together with weather and location information allows a range of possibilities in providing the mobile users with m-adverts that really fulfill their information needs.

Although the number of potential receivers would be considerably smaller, a well-planned execution of m-advertising can be more effective than, for example, direct mail (which is often left unopened) or television advertising (which is restricted by zapping). Based on the personal information, the weather, the time of the day or week, and the location of the consumer, the retailer can send adverts that match with the mobile user's personal interests and mood status. This way, the customers will only receive adverts that they are willing to receive, and any unpleasant surprises are avoided. This is extremely important in permission-based m-advertising since spam messages annoy the receiver (compare Barwise & Strong, 2002; Edwards, Li & Lee, 2002). Therefore, the advertiser can reach high view-through rates by targeting the advert successfully. Research into SMS m-advertising reports that 81% of trialists viewed all messages before deleting them, and 77% did so as soon as they received the advert (Barwise & Strong, 2002). At the same time, this means that the same advert should only be sent to each customer once during a campaign. If the campaign contains repetition, the m-ads have to be different each time they are being sent to the same consumers, otherwise customer annoyance could become an issue.

Advertising Content

As for the content of adverts, the advertiser in any type of advertising has to decide what is being said and how to say it. Both of these decisions affect the success of m-advertising. Kalakota and Robinson (2002) suggest that m-ads work best if customers receive concrete benefits from them, like retail alerts, coupons, special offers, and m-tickets, in other words, something that they can act on. However, Barwise and Strong (2002) bring forward six types of adverts as suitable for permission-based m-advertising, ranging from messages aimed at long-term effects, in other words, brand building, to messages attempting to engage the receiver in immediate interaction with the advertiser, such as competitions, polls, and voting. We suggest that by applying the information given by the consumer and/or information retrieved from the CRM databases, the advertiser can also provide quick and timely information (i.e., news that interests the receiver). So far, with existing research being so scarce, it is too early to say what type of content m-advertising channel suits best.

The style of the advert is also an important issue to be considered. Duchnicky and Kolers (1983) suggest that reading from mobile devices may take more time and effort than reading from a desktop computer. Because of this and also due to the space limitations, the copy should be kept short, and the use of graphics or photos is encouraged (compare Edens & McCormick, 2000). Humor and surprises in the design of the advert create positive feelings toward the advert and may lead to viral marketing, especially amongst younger receivers (Barwise & Strong, 2002). Furthermore, we assume that the personal nature of the mobile devices, as well as the context specificity and novelty of m-advertising will lead consumers toward high involvement. In such situations, the contrast affect appears to stimulate consumers to process the advertising even more (De Pelsmacker, Geuens & Anckaert, 2002). We will now move on to explore the case and to present some empirically grounded suggestions for retailers' using m-advertising.

A Case Study of Permission-Based M-Advertising

The empirical part of the chapter is derived from SmartRotuaari service system, which is operational at the city center of Oulu in Northern Finland. SmartRotuaari

Table 1. Technologies used in SmartRotuaari

Connectivity	WLAN IEEE 802.11b Cisco air ap1220b+adsl
Mobile device	PDA, HP iPAQ 38 series + Pocket PC 2002+WLAN card + suitable mobile phones
Programming environment	Client: Personal Java 1.1.x Server: J2SE 1.4 server, PHP
Protocols	HTTP, SIP, SOAP, XML SQL
Database	MySQL
WLAN positioning	Ekahau Positioning Engine 2.0
Personalization engine	Leiki Targeting
Language support	Dynamic, Finnish and English

consists of a wireless multi-access network, a middleware for service provisioning, a Web portal with retailer interface, and a collection of context aware mobile services of which m-advertising is one (compare Ojala et al., 2003). The retailers use a Web portal to send the adverts, which are delivered through WLAN network to consumers using mobile devices (e.g., PDAs). The technologies that are being applied in SmartRotuaari are presented in Table 1.

Data Gathering and Methods of Analysis

SmartRotuaari provides a functional framework for field trials in order to empirically evaluate technology and new mobile services (*www.rotuarri.net*). The empirical data concerning the retailers' use of permission-based m-advertising presented in this chapter relates to a field trial, which started August 28 and ended September 30, 2003. Altogether, 18 retailers agreed to take part in the field trial, but only 12 were active in designing and sending m-adverts via the service system. Based on personal interviews of all the 18 retailers, the main reasons for not using SmartRotuaari were unfamiliarity with the technology and time restrictions. It would have taken time to learn how the system works and also to design the adverts. This suggests that retailers need training before they would be able to use m-advertising.

The 12 companies that m-advertised (clothing retailers, restaurants, cafeterias, bookstores, and consumer service companies) altogether owned 24 outlets in the area and produced 42 m-adverts during the trial. To see how well the retailers were able to meet the demands of the media and the customers, we will explore the targeting and the content of m-adverts by content analysis.

Content analysis is the standard analytical tool used for advertising studies (compare Kassarjian, 1977; Kolbe & Burnett, 1991). Here, the unit of analysis was each m-advert that was saved to the service system. However, we did not include the number of times the advert had been sent to receivers in the analysis as it might skew the results (Stern & Resnik, 1991). As suggested by Kassarjian (1977), two coders analyzed the commercials. However, due to confidentiality of the data, the authors served as coders. The data was divided into two parts and coders A and B (B being one of the authors) individually coded the first part of the data, which contained the targeting information selected by the advertisers. Coders C and D (D being one of the authors) individually coded the second part of the data, which was the content of the adverts. The authors provided coders A and C with instructions and brief training before they commenced the task. After the first round of coding, there were some disagreements. When this occurred, the coders examined the advert together, discussed the disagreement, and then made their final decision. Since the number of adverts was relatively small, all disagreements between the two pairs of coders were solved through discussion. The aim of this procedure was to achieve higher objectivity (Kassarjian, 1977). As for reliability, since the coders agreed with all the decisions made, no measures of interjudge reliability were calculated (compare Perreault & Leigh, 1989).

The Targeting of the M-Adverts

We will first take a look at the context-related options that advertisers used to trigger adverts. All 42 m-adverts used the advertiser's store location as the focal point from which the distance that triggered the sending of an advert was measured. However, there were huge differences in the ways the retailers used location awareness. The distance used varied from 70 yards up to two miles. The size of the town center in Oulu is under the two miles, so the use of the highest distance in the location awareness does not aid the targeting of the ads.

The second most used feature by retailers in targeting was age. Bars and pubs especially targeted the adverts toward either young or mature customers. Only some clothing shops and a few restaurants left the age option unused.

The mood information was used in 65% of all the advertisements. The clothing shops and cafeterias, especially selected customers that were in the mood for shopping, and bars and restaurants chose people who were hungry, thirsty, seeking company, or in the mood for a party. As for the consumers' interest areas, only 14 of the m-adverts included certain interest areas as criteria for targeting.

Time awareness, according to certain hours of the day (e.g., opening hours, lunch hours), was used in only 18 m-adverts, although it could have been used in every one to ensure that customers would only receive m-adverts during the opening hours. None of the 42 adverts used the local weather as triggering criteria.

It can be concluded that mood information was the specific feature that retailers considered to be worth using in order to target their m-adverts to fit the needs of the consumers. However, they did not use location awareness (i.e., weather or time as targeting criteria), although these options could ensure the m-advert fits in the consumer's current needs. Nor did retailers try to engage consumers in interactive dialog. Thus, instead of personalizing and targeting the m-ads to individual target persons, the retailers used such targeting criteria available in other media.

The Content of the M-Adverts

Another key issue in designing effective m-advertising is the personalization of the content of the m-advert. In this trial, although the majority of the m-adverts (55%) contained visual elements, photos (of people, products, or the interior of a restaurant), or graphics, their use was seldom entertaining, as most of them were static. The copy length ranged from zero to thirty-one words. Although the longest copy was readable on PDAs, the screen was filled with text. This certainly provides no aesthetic pleasure for the targeted individual.

Many adverts (40%) included the address of the store, although it was possible for the consumer to also use a mobile map to locate the company. Only three ads included the phone number of the store. Moreover, 45% of the adverts contained information on opening hours, which explains the fact that many advertisers did not use the option of restricting sending of the advert within opening hours only. They clearly could not use the option, since only a few other media can offer the same. One-third of the adverts included price information or special offers, thus responding to the consumers' information needs. Moreover, only three adverts addressed the receiver in the copy by asking them a question (Are you hungry?) or by welcoming them to the cafe. None of the m-ads attempted to create a dialog with the receiver, not to engage them in viral marketing. Finally, only a single m-advert aimed at brand building. This can be explained by the fact that all the advertisers were retailers, and thus most of the adverts concentrated on describing the shop or the restaurant (e.g., what type of food was served).

Based on this data, retailers and their advertising agencies have to solve the question of how to fit the message and the format into the context of m-advertising (compare Kiani, 1998; Kunoe, 1998). In this empirical trial, m-adverts resembled traditional newspaper adverts that target groups of people. The content of the m-adverts was not well personalized, and only few attempts at interaction with the target person were made. Most of the m-ads offered information of a type that a person living in Oulu had no need for, and only a few had entertaining elements attached to them.

Managerial Recommendations

Permission-based m-advertising is a different and a novel way to perform marketing communication. For advertisers, as well as for advertising agencies, it is crucial to understand that m-advertising is a very personal communication medium that enables them to engage consumers in a dialog. Secondly, if m-advertising is used in a company's media plans, the message and its form should be planned specifically so that it fits the context (i.e., a mobile device). Moreover, the advertiser should consider these decisions, not only when using m-advertising, but in relation to all other media choices to be able to integrate the message, the form, and the media choices.

M-advertising is ideal for one-on-one personal and speedy communication with consumers who wish to receive the information and/or entertainment offered by the retailer. It is very important to harness the power of the context dependence in order fulfill the needs and wants of individual customers. This can be done by applying the options that the m-advertising service system offers to target specific m-ads to specific customers. M-advertising is only one solution to personal, interactive, and context-dependent advertising. There-fore, m-advertising will become fully utilized when managers see m-advertising as an essential part of advertising portfolio that should be included in the future plans and budgets of every company wanting to survive the interlinked m-commerce era.

Future Trends

As the m-advertising business is still evolving, several actors can take part in the network. These include advertising agencies, media houses, media sales companies, network carriers and operators, technology and software providers (compare Leppäniemi, Karjaluoto & Salo, 2004). The traditional ad agencies seem to be reluctant to change from mass communication to personal communication (Suokko, 2003), and thus companies, such as UK-based 12snap and Finland-based Add2Phone, are offering mobile marketing software directly to advertisers. A similar change occurred back when software packages, such as PageMaker, enabled companies to internalize the design of brochures and catalogues, until they realized the value of professional art directors and copywriters.

In addition, the technology platforms for m-ads are likely to evolve into a single platform for multiple purposes in a similar way as PCs. When this happens, advertisers could send similar m-ads to any or all different handheld devices used by consumers. This enables the possibility for advertisers and consumers to determine the preferred route of m-advertising, for example, via a telecommunication carrier, an Internet service provider, or Bluetooth. In addition, the consumers who wish to maximize speed and usability according to their user habits might even wish to pay to be able to take part in dialog marketing (compare Schmidt, 2001). The mobile phone is already a vital part of people's everyday lives, to the extent that many consumers could not cope without it (http://www.kampanja.net). Thus m-advertising should enable interactivity and create feelings of mobile community. Once the advertisers and advertising agencies learn how to use m-advertising, customers will get more value from it (Schmidt, 2001).

Limitations

This study is a part of an ongoing research project where m-advertising was carried out in a WLAN infrastructure with PDAs. The nature of the field test in a limited area and time frame restricted the amount and quality of data. First, the retailers did not pay anything for their m-advertising. Second, the PDAs were not owned by the volunteer customers. The trial time per user had to be restricted to two hours because of the batteries of the PDAs. Consequently, due to the above mentioned restrictions, further studies are warranted.

Conclusions

This study of retailers' usage of permission-based mobile advertising underlies the notion that mobile advertising is different from any other form of advertising, even from direct advertising in its current forms. Thus, both the receivers and the senders of mobile advertising messages have to learn how to use this new channel to be able to fully utilize the opportunities it offers for speedy, personal, and interactive communication with the consumer (compare Pura, 2002).

M-advertising should be personal, thus requiring a certain amount of knowledge about the receivers of the m-adverts. This can be achieved either by using a company's own customer information data or providing each phone user an opportunity to fill in a profile when agreeing to receive m-adverts. The message of the advert as well as the way it is expressed should be carefully designed to match the needs of the target person. M-adverts containing more information should be directed to persons using their mobile devices for that goal and entertaining m-adverts to youngsters that entertain themselves with the device (e.g., by playing games).

Moreover, the advertisers should design m-advertising to fit their marketing communication mix, enabling interactivity. Each customer should be able to use the mobile device to contact the retailer via Internet by simply sending an SMS or phoning. In addition, customers should be encouraged in viral marketing by offering them added value, either monetary or entertainment, that they want to share with their friends and family.

Acknowledgment

The financial support of the National Technology Agency of Finland is gratefully acknowledged. The authors wish to thank the numerous organizations whose invaluable collaboration has made this work possible.

References

Anastasi, G., Bandelloni, R., Conti, M., Delmastro, F., Gregori, E., & Mainetto, G. (2003). Experimenting an indoor Bluetooth-based positioning service. *Proceedings of the International Conference on Distributed Computing Systems Workshops*, *23*, 480–483.

Barwise, P., & Strong, C. (2002). Permission-based mobile advertising. *Journal of Interactive Marketing, 16*, 14-24.

Choi, S.Y., Stahl, D.O., & Whinston, A.B. (1997). *The economics of electronic commerce*. Indianapolis: Macmillan Technical.

Clarke, I., III, & Flaherty, T.B. (2003). Mobile portals: The development of m-commerce gateways. In B.E. Mennecke & T.J. Strader (Eds.), *Mobile commerce:Technology, theory and applications* (pp. 185-201). Hershey, PA: Idea Group.

De Pelsmacker, P., Geuens, M., & Anckaert, P. (2002). Media context and advertising effectiveness: The role of context appreciation and context/ad similarity. *Journal of Advertising, 31*, 49-61.

Durlacher Research Ltd. (2000). *UMTS report: An investment perspective*. Retrieved October 14, 2004, from *http://www.durlacher.com*

Edens, K.M., & McCormick, C.B. (2000). How do adolescents process advertisements? The influence of ad characteristics, processing objective, and gender. *Contemporary Educational Psychology, 25*, 450–463.

Edwards, S.M., Li, H., & Lee, J-H. (2002). Forced exposure and psychological reactance: Antecedents and consequences of the perceived intrusiveness of pop-up ads. *Journal of Advertising, 31*, 83–95.

Eysenck, M. (1982). *Attention and arousal, cognition and performance*. New York: Springer-Verlag.

Godin, S. (1999). *Permission marketing*. New York: Simon & Schuster.

Goldsborough, R. (1995, May 8). Hong Kong trams keep ads rolling. *Advertising Age, 66*, 36.

Hume, S. (1988, April 11). New medium is semi success. *Advertising Age, 59*, 22-24.

Juntunen, A. (2001). *Audience members' goals of media use and processing of advertisements*. Acta Universitatis Oeconomica Helsingiensis A-187, Helsinki School of Economics and Business Administration, Helsinki, Finland.

Kalakota, R., & Robinson, M. (2002). M-business. *The race to mobility.* New York: McGraw-Hill.

Kassarjian, H. (1977). Content analysis in consumer research. *Journal of Consumer Research, 4,* 8-18.

Kiani, G.R. (1998). Marketing opportunities in the digital world. *Internet Research: Electronic Networking Applications and Policy, 8,* 185–194.

Kotanen, A., Hännikäinen, M., Leppäkoski, H., & Hämäläinen, T. (2003, April 28-30). Experiments on local positioning with Bluetooth. *Proceedings of International Conference on Information Technology: Computers and Communications,* 297–303. *http://csdl.computer.org/ comp/pro0ceedings/itcc/2003/1916/00/19160297abs.htm>http:// csdl.com*

Kunoe, G. (1998). On the ability of ad agencies to assist in developing one-to-one communication. Measuring "the core dialogue". *European Journal of Marketing, 32,* 1124–1137.

Leppäniemi, M., Karjaluoto, H., & Salo, J. (2004). The success factors of mobile advertising value chain. *E-business Review, 4,* 93-97.

Liu, Y., & Shrum, L.J. (2002). What is interactivity and is it always such a good thing? Implications of definition, person, and situation for the influence of interactivity on advertising effectiveness. *Journal of Marketing, 31,* 53–64.

Mennecke, B.E., & Strader, T.J. (2003). *Mobile commerce: Technology, theory and applications.* Hershey, PA: Idea Group.

Ministry of Transport and Communications Finland. (2003). *Ensimmäisen aallon harjalla. Tekstiviesti-, WAP- ja MMS-palveluiden markkinat 2000–2004.* [On the first wave]. Publication of the Ministry of Transport and Communications Finland. Retrieved October 14, 2004, from *http:/ /www.mintc.fi/www/sivut/dokumentit/julkaisu/julkaisusarja/2003/ a192003.pdf*

Ojala, T., Korhonen, M., Aittola, M., Ollila, M., Koivumaki, T., & Tahiten, J. (2003, December 10-12). SmartRotuaari – Context-aware mobile multimedia services. *Proceedings of the Second International Conference on Mobile and Ubiquitous Multimedia,* 9-18. *Norrköping, Sweden. http://www.ep.liu.se/ecp/011/005/*

Perreault, W.D., & Leigh, L. (1989). Reliability of nominal data base on qualitative judgments. *Journal of Marketing Research, 26,* 135-148.

Pervin, L. (1989). Goals concepts: Themes, issues, and questions. In L.A. Pervin (Ed.), *Goal Concepts in personality and social psychology* (pp. 473-479). Hillsdale, NJ: Lawrence Erlbaum.

Pura, M. (2002). Case study: The role of mobile advertising in building a brand. In B.E. Mennecke & T.J. Strader (Eds.), *Mobile Commerce: Technology, Theory and Applications* (pp. 291-308). Hershey, PA: Idea Group.

Schmidt, S.J. (2001). Advertising in search of its future. *Poetics, 29,* 283-293.

Stern, B.L., & Resnik, A.J. (1991). Information content in television advertising, a replication and extension. *Journal of Advertising Research, 30,* 36-46.

Suokko, T. (2003). *Markkinointiviestinnän lapsuuden loppu.* Juva: WSOY.

Turban, E., King, D., Lee, J., Warkentin, M., & Chung, H.M. (2002). *Electronic commerce: A managerial perspective.* Upper Saddle River, NJ: Prentice Hall.

Endnotes

[1]   There are several technologies that enable context dependency, that is, global positioning systems (GPS), geographical information systems (GIS), satellite-based wireless systems, and cell identity based systems (Kotanen et al., 2003), or it can be achieved by triangulating the position of the device (Anastasi et al., 2003).

Section III:

Technology for E-Marketing

Chapter IX

Integrating Internet/Database Marketing for CRM

Sally Rao, Adelaide University, Australia

Chris O'Leary, MSI Business Systems Pty Ltd, Australia

Abstract

Firms have only just begun to fully use the Internet to obtain customer information in their database marketing processes to enhance customer relationship management (CRM). This chapter introduces a framework about how they can do this. Essentially, it argues that the advent of Internet/database marketing brings solutions to some of the difficulties in customer relationship management by providing one-to-one interactivity and customization. For example, the Internet offers benefits, such as increased consumer data collection accuracy and speed, cost savings in collecting data, greater interaction, and better relationships with customers. This chapter develops a framework for integrating the Internet and database marketing to help marketers improve customer relationship management through rigorous action research.

Background

The growth in database marketing and the emergence of e-commerce driven by the exponential growth of the Internet requires marketers to capitalize on the full advantage provided by information technology to be competitive. The key component of database marketing is its ability to enhance an organization's marketing program by identifying customers that are likely to be more receptive to a specific offering. Indeed, competent database marketing practice needs to be integrated with other marketing strategies and practices. The interactive Web environment and the advent of Internet marketing present an explicit opportunity for firms to achieve maximum database marketing benefits.

Although the basic database marketing principles are the same, the integration of the Internet into database marketing process allows personalized interaction and communication. That is, the Web medium enables the capture and use of highly personalized information, such as name, interest, type of car owned, TV programs watched, and so on. This information, in turn, facilitates one-to-one marketing (Gillenson, Sherrell & Chen, 1999). For example, armed with customer transaction data and/or third party lifestyle data from companies like Claritas and Acxiom, online direct marketers can deliver personalized interactive promotions, realizing full capabilities of the Web.

A prerequisite for the successful translation of the relationship marketing paradigm from industrial to consumer markets is accurate customer information. Improved quality of customer information enables marketers to target their most valuable prospects more effectively, tailor their offerings to individual needs, improve customer satisfaction and retention, and identify opportunities for new products or services. Therefore, the key focus of e-marketing is customer data that can be used to inform operational, tactical, and strategic decision making (Chaffey et al., 2003).

The Internet offers a valuable opportunity to collect information about a customer (Rowley, 1999). Through the Internet, every customer contact can be used as an opportunity to collect data about the customer, which in turn is incorporated into the marketing database to develop completed customers profiles. Such information can then be matched to numerous databases internal or external to the organization, yielding rich permutations of consumer profiles at a minimal cost. This combined data can then be used to build customer knowledge which drives marketing strategy to develop practical long-term relationships with customers. These relationships, if successful, will lead to further detailed customer-provided data and/or sales that yield more data and therefore start the circle over again (Hughes, 1991). The two-way interactive nature of the Internet mechanisms presents an opportunity for firms to increase the speed and volume of this circle of customer data collection.

In brief, Internet technology has made it relatively easy to collect vast amounts of individual customer information (Prabhaker, 2000). Data quality, entity recognition, synchronization, and integrated databases enable firms to target content to demographics so precisely that they can reach markets as narrowly defined as a single customer. This results in greater levels of customer satisfaction and increased organizational learning. However, merging the off-line and online databases raises issues, such as technology compatibility, data quality and format variances, how to use the sheer volume of data collected, and consumer privacy concerns. This chapter aims to develop a framework about how the integration of database and Internet marketing can be applied, based on rigorous research.

Building a Theoretical Framework Using an Action Research Methodology

An action research methodology was adopted to build a framework about integrating Internet and database marketing. Action research is a cyclical process methodology that incorporates the processes of planning, acting, observing, and reflecting on results generated from a particular project or body of work (Dick, 2000; Zuber-Skerritt & Perry, 2000). The concept is principally concerned with a group of people who work together to improve their work processes (Carson et al., 2001), such as the identification of apposite strategic outputs from the integration of Internet/database marketing to support e-marketing.

The choice of action research as the methodology was based on two factors. First, due to the minimal research that has been conducted regarding the appropriate process for integrating customer information for e-marketing, the manner through which this may be effectively completed was unclear. Thus, exploratory research was required, and action research provides this capability better than many alternatives (Dick, 2000; Zuber-Skerritt & Perry, 2000). An action research project within a major multinational entertainment institution was used to discern the issues involved. The second reason was the flexibility afforded by action research, which was significant in the implementation of a research methodology within an evolving information technology project concerning a problem about which little was known (Carson et al., 2001).

The action research methodology commences with a group jointly concerned with a successful conclusion doing fact-finding about a particular problem in which existing research and understanding is minimal (Altrichter et al., 2000; Edwards & Bruce, 2000). This group in the researched multinational entertainment company involved a project team concerned with the integration of customers from across various entertainment divisions and used customers' explicit and implicit interests to drive a customer-centric e-marketing strategy. The company's divisions are vast, including free-to-air and cable television, print, Internet portals, gambling, sports, show entertainment, and so on. The project team consisted of middle-to-senior management of the entertainment company in addition to the researchers; the joint goal was how to best collect customer data through the Internet, the integration of this data with on and off-line databases, and the best e-marketing strategies to implement.

The project team agreed that Internet and database marketing offers the company a great opportunity to gather market intelligence, monitor consumer choices, and achieve a closer client relationship through customers' revealed preferences in navigational and purchasing behavior. An initial plan for online data collection, integration with off-line data, and preliminary marketing strategies to implement was formulated (Edwards & Bruce, 2000). In particular, a framework outlining the process through which Internet-derived customer data may be effectively integrated with other customer data to enable competent database marketing was discussed and modeled in Figure 1. The framework has the three usual parts of a system: inputs, processes, and outputs. That is, inputs of information about customers gathered via the Internet are processed and converged with a firm's customer databases to produce a series of strategic outputs. Consider these three parts in more detail.

Figure 1. Initial theoretical framework for integration

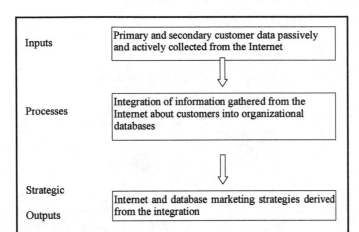

Input

A logical place for the entertainment company to commence collecting customer information through the Internet was their established customer databases derived from more traditional customer contact, billing, and marketing mechanisms. The main customer data categories that could be collected both online and off-line for competent database marketing were listed in Table 1 after a brainstorm session.

The project team agreed that for the conduct of effective e-marketing, the data items in Table 1 need to be collected at a spatial and/or longitudinal detail. This detail allowed the customer to be effectively recognized at each transaction, and for a profile of the customer to be increasingly developed and enhanced over time based on their response history to online content, expressed preferences and previous offers (Seybold, 1999). This approach, in combination with using cross-promotional offers of its other entertainment divisions and offerings, was designed to enable the entertainment company to progressively realize the benefits of e-marketing. The development of this customer profile and associated interest/purchasing patterns were broadened and expedited by utilizing the following:

- modeling of customer relationships, scenarios, and propensities using investigation models, such as regression and interest profiling based on online behavior;

Table 1. Main customer data categories collected for marketing

Primary Data Category	Example Data Elements
Recognition data	Customer unique identifier Name Address (home, work, phone, e-mail) Gender Date of birth
Descriptive data	Income Affluence Total investments Marital status Children (number, age, gender)
Transaction history	Amounts spent Vendor Purchase category Date Channel (online)
Direct preference measures	Purchase detail (product, vendor attributes) Provided preferences and permissions Response history (to types of marketing messages and product offers) Measure of media used Questionnaires (online and off-line) Cookies (determining a specific user or a return user and measuring time spent and where at site and entry and exit points)
Trigger events	Indicators of life events (relocation, college graduation, child birth, saving deposit for house) Measures of product information browsing or requests Cluster product purchase event (for example, home purchase drives need for home and contents insurance) Direct client interest (client informs company of interest in purchasing product/service)

- data mining and online analytical processing (OLAP) to elicit evolving patterns of usage, trends, and customer life-cycle development;
- determination of consumer financial value to the organization through measurement against profitability archetypes, such as recency, frequency, and monetary purchase values (RFM analysis) and customer tenure, derived revenue, and cross-sell opportunities (TRC categorization);

- visualization of data through graphics to make more apparent trends and tendencies; and

- artificial intelligence, such as neural networks and complex adaptive systems theory.

The output from these methods is then integrated back into the data components outlined in Table 1 to provide an ongoing and evolving customer profile that is further enhanced by integrating auxiliary customer interaction and data collection. The integration process is complicated however by the often unstructured nature of data collected through the Internet, the diverse nature, quality, and format of existing company customer data, and the large volumes of data to handle in an e-marketing context. The details of this integration process are discussed next.

Processes

The information gathered from the Internet was processed through identification, standardization, de-duplication, and consolidation procedures, and a unique persistent database key was applied to each individual customer (Inmon, 2002). Records were stored in an organizational data warehouse and updated automatically through data derivation methods, such as Web form logins (for identification) and cookies (to track interests). Figure 2 shows the process through which customer data were integrated and then used for e-marketing. The integration processes in Figure 2 can be summarized into three steps:

Step one: data content identification and understanding. The first step of data integration is data identification and understanding of utilized data content. This step involves the extraction of data feeds in structured format (for example, in XML or comma/pipe separated values) from the company's legacy systems and each of the disparate data sources, such as existing customer data, public data, and purchased data collected by a third party. This data was then subjected by the project team to a data quality audit, leading to a full understanding of the requisite data attributes. A standard data definition and associated metadata standard was hereafter developed for the applicable data attributes, and the e-marketing data elements were prepared for standardization in this format (Corey et. al., 1998; Dyché, McKnight & Adelman, 2003).

Figure 2. Integration of customer data from disparate sources for e-marketing

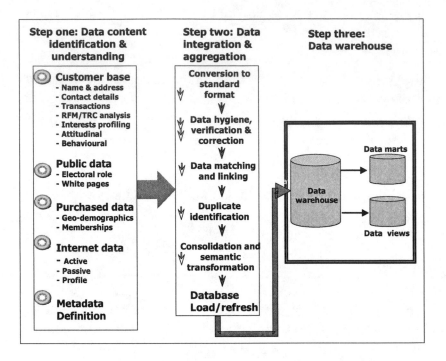

Step two: Data integration and data aggregation. In this second step, data from the company's internal systems, as well as utilized external data was converted into a standard format using the metadata and conversion standards developed in step one. Following this conversion, data attribute correction and verification were undertaken, wherein invalid characters/records were flagged, errors were corrected, and important contact/recognition attributes, such as address and e-mail, were verified (Canter, 2002; Todman, 2000).

The next phase, data matching and linking, involved the company using specialized software and associated processes to identify duplicate records and logically joining these through a common persistent identifier or link (English, 1999; Todman, 2000). The matching utilized by the entertainment company consisted of two iterations:

1.   the initial progression, implemented at the original data warehouse data load, which was effected as a batch process; and

2.   ongoing updates, or deltas, which were a combination of real-time and batch processes.

Following the matching and linking process, de-duplication was performed and the data consolidated. From here, semantic transformations, such as address attribute reformating to facilitate the data load process into the data warehouse was effected, with particular attention paid by the project team to effect these transformations in real-time to allow the system to cope with its continuous rapid data updates. At this point, the consolidated data is ready for loading into the data warehouse repository.

Step three: The data warehouse. Following completion of steps one and two, the cleansed de-duplicated data was loaded into the central data warehouse. The data warehouse is an enterprise data construct that integrates, stores, and maintains customer data derived from organizational legacy systems and external sources and acts as the cornerstone for e-marketing strategies and operations through its focus on customer related information (Van Dyk, 2002). This data repository was large, and therefore, the project team decided to create data marts—specific data views or subsets of the data warehouse optimized to a particular organizational division's needs to assist users in accessing this information.

Strategic Output

The final step involves the implementation of e-marketing applications to drive strategic output. Customer data collected through the Internet was used in such application as:

- reporting and modeling applications to facilitate customer understanding and identifying interests, trends, and propensities;
- campaign management, to distinguish customers according to likelihood of responding to a specific campaign initiative, disseminate the campaign via e-mail or customized Web pages (refer below) and collect and report feedback; and
- real-time analytical/content personalization, which was used to predict customer interests based on provided information and Web page click stream or navigation behavior and use these interests to deliver personalized Web content and e-marketing offers.

The project team agreed that integrated customer data collected from online and off-line sources needs to be used more strategically than just as a tool for conducting targeted promotional campaigns. After a brainstorming session, four e-marketing strategies were identified most likely to be derived from the integration of Internet and database marketing to improve customer relationship management and ultimately implemented. These strategies are:

1. *Prospecting new customers through affiliate or "stealth" marketing and networking*: The wide range of entertainment avenues and associated product/service offerings provided by the company, in combination with those of partners and advertisers, provided a compelling Web site, which lured many prospective customers through a combination of off-line brand, special offers, and rewards for registering/customer online identification and interest specification. For example, the entertainment company incorporated Web content with popular television shows and magazines also owned by the entertainment company and offered holidays, subscription savings, and special interest related purchases via this Web content in return for the customer providing personal details and effecting purchases. This combination of offering off-line brands through a data-driven Internet marketing strategy proved successful, with the new Web site growing rapidly to become one of Australia's most popularly visited.

2. *Promoting and advertising pertinent to a customer's identified interests:* The strategy of affiliate marketing and rewarding customer identification and interest specification allowed the company to structure promotion and advertising to be more closely aligned with the customer's real interests. This alignment was evidenced by increased network traffic and uptake of offered advertising. An example is the entertainment company structuring its advertising to be associated and driven by the specific Web pages in which customers were browsing, in that specific sports promotions were associated with specific sports pages, such as cricket, rugby or golf, holidays with lifestyle travel Web pages, and so on.

3. *Cross-marketing*: Further to advertising according to a customer's interests, as discussed above, the data integration of Internet customer data with legacy/external information allowed the entertainment organization to affect cross-marketing initiatives. The entertainment company, for example, used online consumer data integrated with off-line magazine subscription information to drive special promotions for lifestyle offers

associated with subscribed magazines through e-marketing. Further, the defined interests provided by customers via the Internet, in combination with the customer data profiling/analysis tools discussed previously, allowed for express interests to be extrapolated into cross-sales opportunities.

4.   *Retaining customers*: The project team determined that customer retention could be achieved by the improved customer understanding and service factors derived from an integrated Internet/database marketing solution. Thus, the entertainment company would utilize initiatives, such as owning print media subscription discounts, related TV and sports special offers, and holiday promotions linked with their reality/lifestyle TV shows that were available only to members through its e-marketing channel. Again, based on customer sign-up, repeat visits, and repeat purchases through the Internet media, this strategy was also deemed effective by the project team.

Managerial Recommendations

Following the implementation of an integrated Internet and database marketing solution in the entertainment company, results were monitored against problem-solving expectations and the efficacy of the actions on the problem were evaluated (Farquhar, 2000). Several implemental issues were identified in this evaluation process. First, feedback from customers during and after the implementation of marketing strategies helped the company to structure its e-marketing mechanisms to assist in further data collection and customer understanding. For example, community/special interest groups, television show fan groups, and so on were used by the entertainment company to encourage customers to more actively contribute their information. Thus, a feedback loop is depicted in the framework (Figure 3).

Second, the evaluation process highlighted several environmental issues at both macro and micro levels regarding the successful integration of Internet and database marketing. At the macro level, social, legal, and technological issues need to be taken into consideration. An important social issue is related to consumer's concern with privacy. There is a delicate balance between the benefits consumers enjoy from personalized customer service and online experience provided from personal information exchange and the amount of information that consumers are prepared for companies to hold about them. The entertainment company was particularly concerned that privacy management be optimally managed given its high profile in order to avoid any negative press and to maximize customer uptake. The project team thus determined that privacy management was integral to the successful integration of the Internet with database marketing.

This also brings up legal issues since many governments have introduced specific data protection legislation to protect consumers, such as the Australian Privacy Act 1999. Meeting privacy legislation and guidelines remains a challenge. The entertainment company addressed these guidelines by adopting a permissions-based transparent disclosure approach. These principles were summarized as a result of this approach:

• Limit the collection of personal data to only that perceived as necessary by the customer;

• The purpose for which data are collected and used must be disclosed to customers;

Figure 3. The final framework for the integration of database marketing into Internet marketing (Source: Developed from this research)

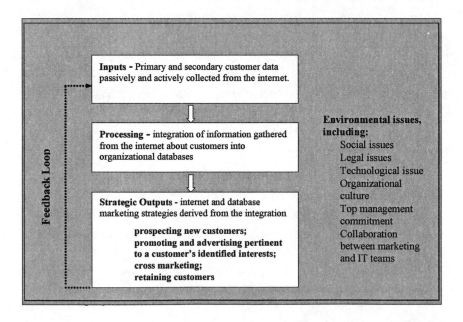

- Consumers must be given options with respect to whether and how personal information may be collected from them;
- Provide customer access to view and validate the accuracy and completeness of data collected about them. As discussed previously, the project team also determined that this factor is effective in maximizing data quality; and
- Safeguard consumer data and place restrictions on the transborder flow of consumer information.

Into the picture also has come the use of new network technologies and techniques which raise new measurement issues. An important example is the extension of e-marketing towards other media, such as mobile phone small text messaging (SMS) and Internet-enabled wireless application protocol (WAP) phones that herald the extension of m-commerce. These new customer interfaces, in combination with the online and off-line customer contact points, form a diverse customer communication base and may present a hindrance to the process of data integration and management.

The entertainment company, organizational culture, top management support, and collaboration between marketing and IT teams provide a supportive environment for the integration and is critical for overall success. First, organizational culture must support innovative use of the Internet in their marketing activities. Second, top management must be philosophically and behaviorally committed to the notion that the Internet is essential in collecting customer data in enabling effective database marketing. Finally, an effective cross-functional team of information systems and marketing specialists must work harmoniously to succeed in the integration of Internet and database marketing.

Future Trends

Increased Usage of Internet and Database Marketing Strategies

As the Internet becomes more pervasive and e-business extends further across industry models and customer access channels, it is expected that the use of database-driven Internet marketing will increase and will become increasingly sophisticated (Sweiger et al., 2002). Further, this increased usage of the Internet is expected to increasingly involve non-Western economies and, in particular, the Chinese and Asian economies (Anon, 2003). As such, marketers may need to adopt multinational, multicultural, and multilingual Internet/database marketing strategies to gain maximum value from this anticipated growth. The experience of the entertainment company confirms these predictions, in that the success of its customer data-driven Internet marketing strategy has resulted in future projects for extension of the concept into South East Asia.

Increased Emphasis on Data Quality and Data Integration

Data quality and effective data integration will continue to be a focus for many firms as building up accurate customer/prospect profiles, interests, and propensities over time is the critical success factor to effectively achieve and maximize the strategic outputs identified in Figure 3. The success of the entertainment company's integrated Internet marketing and database marketing strategy (as measure by increased site traffic, numbers of customers signing on, interest-generated responses to offers, and increase in total sales) has resulted from the company being able to maintain data quality. The entertainment company encouraged customers to identify themselves as quickly as possible in return for compelling content and special offers. A persistent customer log-on was assigned to the customer, and rewards, such as product discounts for customer validation and upkeep of their own profile and use of external reputable data for validation of attributes such as Internet domain name, address, and phone details, were used to increase data quality.

As such, as integrated Internet and database marketing becomes more mainstream within the entertainment company, failure to adequately address the issues of data quality, customer recognition, and data integration, especially in a real-time capacity, will increasingly negate the efficacy of this profile building and thus frustrate the ability to deliver the strategic outputs. Further, as competitors increasingly attempt to replicate the success of the data-driven Internet marketing strategies enjoyed by the entertainment company, it is believed by the project team that organizations that fail to effectively monitor data quality and integration procedures risk losing market share to competitors that can successfully build on and utilize their prospect/customer information assets. The future will place greater reliance on data quality and the associated intelligent use of integrated information.

Prominence on "Value-For-Money" and Differentiation Strategies

As integrated Internet/database marketing strategies become more commercially utilized, increased emphasis on innovative differentiation and "value-for-money" across the identified strategies is expected to emerge. The future success in e-marketing comes when online promotions and catalogues are delivered fully personalized to meet the needs of each individual customer. As the prospect/customer becomes increasingly inundated with requests for information and data-driven Internet marketing offers, expected future trends will involve an enhanced expectation for incentives and/or rewards (financial, prestige, status, other) for the targeted persons to contribute to and keep their e-marketing data current.

Increased Emphasis on Privacy Management

The increased Internet usage and customer expectations discussed above are expected to result in increased emphasis on privacy and customer permissions management (Godin, 1999). This increased expectation for information recompense will be accompanied by increased customer/prospect intolerance for irrelevant or inappropriate offers and a corresponding increased emphasis on privacy (Hagel & Singer, 1999). Privacy problems are difficult and often solved on an ad hoc basis. Although self-regulation is the most common approach in handling consumer privacy protection, more stringent government regulation will demand compliance in the future. For example, the Australian Federal Government imposed a stricter SPAM Act recently to regulate unsolicited commercial messages. As such, organizations are expected to become increasingly sophisticated in setting, tracking, and managing privacy across their Internet marketing strategies. Technological innovation, such as secure e-mail services, anonymous surfing tools, and cookie cutters, are also becoming available which present greater challenges to marketers to collect customer information.

Conclusions

In conclusion, the extent, degree. and speed of communication enabled by the Internet makes it a synergistic component of an effectual database marketing strategy. The assimilation of all the marketing tools, including the two powerful ones – the Internet and databases built on a platform of comprehensive data integration – is the key to executing effective one-to-one marketing. An integrated marketing solution appears promising in the development of e-marketing. An action research approach was used to build a framework about how Internet marketing and database marketing can be integrated. The results are qualitative in nature and further research is needed to establish the statistical generalizability of these theory-building findings in survey research when the population of marketing managers with the requisite expertise becomes large enough for a survey to be done.

References

Altrichter, H., Kemmis, S., McTaggart, R., & Zuber-Skerritt, O. (2000). The concept of action research. In *Action learning, action research and process management: Theory, practice, praxis* (pp. 84-98). Brisbane: Griffith University, Action Research Unit.

Anonymous. (2003). Asian Web users. Retrieved October 15, 2004, from *http://www.pegasusresearch.net/metrics/growthas.htm*

Canter, J. (2002). Today's data warehouse demands quality data. *Journal of Data Warehousing, 7*(2), 56-62.

Carson, D., Gilmore, A., Gronhaug, K., & Perry, C. (2001). *Qualitative research in marketing.* London: Sage.

Chaffey, D., Mayer, R., Johnston, K., & Ellis-Chadwick, F. (2003). *Internet marketing – Strategy, implementation and practice* (2nd ed.). London: Prentice Hall.

Corey, M.J., Abbey, M., Abramson, I., & Taub, B. (1998). *Oracle 8 data warehousing: A practical guide to successful data warehouse analysis, build and rollout.* Berkeley, CA: Osborne/McGraw-Hill.

Dick, R. (2000). Postgraduate programs using action research. In *Action learning, action research and process management: Theory, practice, praxis* (pp. 102-138). Brisbane: Griffith University, Action Research Unit.

Dyché, J., McKnight, W., & Adelman, S. (2003). Experts' perspective. *Business Intelligence Journal, 8*(2), 34-39.

Edwards, S.L., & Bruce, C. (2000). Reflective Internet searching: An action research model. In *Action learning, action research and process management: Theory, practice, praxis* (pp. 180-188). Brisbane: Griffith University, Action Research Unit.

English, L.P. (1999). *Improving data warehouse and business information quality: Methods for reducing costs and increasing profits.* New York: John Wiley & Sons.

Farquhar, M. (2000). Reflections of a founding member of ALARPM: An interview with Ortun Zuber-Skerritt. In *Action learning, action research and process management: Theory, practice, praxis* (pp. 189-204). Brisbane: Griffith University, Action Research Unit.

Frawley, A., & Thearling, K. (1999). Increasing customer value by integrating data mining and campaign management software. *Direct Marketing, 61*(10), 49-54.

Gillenson, M., Sherrell, D., & Chen, L. (1999). Information technology as the enabler of one-to-one marketing. *Communications of the Association for Information Systems, 2.* Retrieved October 15, 2004, from *http://cais.isworld.org/articles/2-18/article.htm*

Godin, S. (1999). *Permission marketing.* New York: Simon & Schuster.

Hagel, J., & Singer, M. (1999). *Net worth: Shaping markets when customers make the rules.* Boston: Harvard Business School Press.

Hughes, A.M. (1991). *The complete database marketer: Tapping your customer base to maximize sales and increase profits.* Chicago: Probus Publishing.

Inmon, W.H. (2002). *Building the data warehouse* (3rd ed.). New York: John Wiley & Sons.

Prabhaker, P.R. (2000). Who owns the online consumer? *Journal of Consumer Marketing, 17*(2), 158-171.

Revans, R.W. (1991). The concept, origin and growth of action learning. In O. Zuber-Skerritt (Ed.), *Action learning, action research and process management*. Brisbane: ALARPM/AEBIS Publishing.

Rowley, J. (1999). Loyalty, the Internet and the weather: the changing nature of marketing information systems? *Management Decision. 37*(6), 514-518.

Seybold, P.B. (1999). *Customers.com: How to create a profitable business strategy for the Internet and beyond.* New York: Times Business.

Sweiger, M., Madsen, M.R., Langston, J., & Lombard, H. (2002). *Clickstream data warehousing.* New York: John Wiley & Sons.

Todman, C. (2000). *Designing a data warehouse: Supporting customer relationship management.* Englewood Cliffs, NJ: Prentice Hall PTR.

Van Dyk, W. (2002). Designing multipurpose data warehouse models. *Journal of Data Warehousing, 7*(3), 31-35.

Zuber-Skerritt, O., & Perry, C. (2000). Action research in graduate management theses. In *Action learning, action research and process management: Theory, practice, praxis* (pp. 189-204). Brisbane: Griffith University, Action Research Unit.

Chapter X

Developing Brand Assets with Wireless Devices

Jari H. Helenius, Swedish School of Economics and Business Administration, Finland

Veronica Liljander, Swedish School of Economics and Business Administration, Finland

Abstract

Advancements of the wired Internet and mobile telecommunications offer companies new opportunities for branding but also create a need to develop the literature to incorporate the new communication channels. This chapter focuses on the mobile channel and how mobile phones can be used in branding activities. Based on a literature review and practical examples, the chapter discusses how brand managers can utilize the mobile channel to strengthen brand assets. Four mobile branding (m-

branding) techniques are proposed and their impact on brand assets discussed. Managerial implications and suggestions for further research are provided.

Introduction

The latter part of the 20th century and the beginning of the 21st will be remembered for the rapid development of consumer communication devices and self-service technologies, most notably the wired Internet, or the Web, and mobile telecommunications. Improvements in the form of faster connections, cheaper usage, and sophisticated hardware in combination with user friendliness have made the Web an increasingly important medium for consumer interaction. Consequently, leading marketing practitioners and researchers specialized in branding have pointed out the need for developing the existing branding frameworks to accommodate the online world (Aaker & Joachimsthaler, 2000). In this development, it is essential to include mobile technologies, most notably the mobile phone, that already offers marketers opportunities for branding. In the future, when the Internet and mobile technologies have fully converged, the possibilities for marketers will be even greater.

The use of mobile devices in marketing and particularly in branding is still in its infancy. Although studies of mobile services are starting to emerge (e.g., Barnes & Corbitt, 2003; Heinonen, 2004; Nordman & Liljander, 2004; Repo, Hyvönen, Panzar & Timonen, 2004), there is still a lack of literature on marketing activities enabled by mobile technology (Balasubramanian, Peterson & Jarvenpaa, 2002; Pura, 2003). Furthermore, to our knowledge, the use of mobile handsets for branding activities has not been the subject of research. In this chapter, we address this gap by discussing the possibilities for mobile branding (m-branding) that are available to marketers today. The focus is on how brand assets can be developed through the use of wireless devices, which in this chapter are limited to cover only mobile phones. At present, the mobile phone is the most ubiquitously used mobile device; it is estimated that there are one billion in use worldwide.

Marketing professionals, however, have not yet adapted their brand strategies to the new technology. The use of mobile phones in marketing has been characterized by trial-and-error activities and has often been met by annoyance reactions from consumers. The slow adoption of mobile technology in branding may be due to the high penetration of low-end mobile devices without color displays, small screens, and limited processing capability, offering primitive and unfriendly brand experiences, which at worst can detract from the brand's value, its brand equity. If the predictions of the future penetration of high-end advanced mobile devices are met, this may soon change. For example, camera phones have been a market success (Crocket & Reinhardt, 2003), and leading telecommunications industry analysts predict that by 2007 some 298 million mobile phones worldwide will be integrated camera phones (*http:// www.idc.com*). With color displays and other improvements, better user experiences can be delivered. However, also less advanced phones can be, and are, used for branding activities.

In Finland, especially among the young, consumers actively use their mobile phones for text messaging, downloading ring-tones, picture logos, and mobile screensavers, ordering mobile horoscopes, playing mobile games, and participating in mobile chat rooms. The mobile phone can be considered a device that is always present and turned on. A personal observation by one of the authors is that, also in Japan, the mobile phone, particularly the I-mode service, forms a permanent presence in consumers' everyday life. For Japanese consumers, using I-mode services has become a daily routine, including reading the news, weather reports and cartoons, performing banking activities (e.g., checking account balances), m-commerce (e.g., reserving concert, movie, or airline tickets), e-mailing, and playing games. These examples demonstrate the growing potential of the mobile channel for marketers.

This chapter discusses the capabilities and opportunities of m-branding that exist today, as well as its current limitations. The chapter is structured as follows. First, brand equity and brand assets are briefly introduced and ways of using the mobile phone in branding activities are discussed. Second, the m-branding methods that are available today are presented with examples of how they can be used. Third, managerial recommendations and future trends of m-branding are discussed.

Brand Equity

In the 1990s, managers increasingly realized that brands represented real value for the company and that they needed to pay more attention to the development of brand equity. Several definitions and descriptions of brand equity can be found in the literature (e.g., Aaker, 1991, 1996; Aaker & Joachmisthaler, 2000; Kapferer, 1997; Keller, 1998), but in summary, it can be concluded that brand equity is the measured outcome of all branding activities and that it can be expressed in financial terms. It is the (financial) value of the brand, based on a number of brand assets. From a brand manager's point of view, it is essential that the brand assets can be managed.

Brand Assets

Aaker and Joachimsthaler (2000) divide brand equity into four contributing factors, or brand assets: brand awareness, brand associations, perceived quality, and brand loyalty. The dimensions are not independent. For example, brand awareness and perceived quality affect brand associations. Brand associations include all the things that consumers connect with the brand, such as the brand personality, product attributes, and symbols. The perceived quality of a brand is believed to have a direct effect on brand profitability, but it can also be expected to have a direct effect on brand loyalty (i.e., the emotional attachment that consumers have to a brand and their repeat purchasing behavior). The stronger and more positive the brand assets are, the higher the brand equity.

Since consumer loyalty has, in general, been strongly associated with positive economic consequences for the company (Reichheld, 1996; Storbacka, Strandvik, & Gronroos 1994), loyalty can be considered the company's most important asset. Defending or increasing brand loyalty should be one of the goals of m-branding activities. Wireless devices offer unique consumer values, such as localization, timeliness, and convenience (Heinonen, 2004; Lindstrom, 2001), which can be used to form and enhance brand associations and thereby loyalty. However, in the absence of long-term studies, the existence of these relationships can only be assumed. It should be noted that although increased brand loyalty is one of the goals of mobile marketing, short-term campaign results, in that regard, have been disappointing (Pura, 2003).

The assets need to be communicated to customers in the form of a coherent brand identity. In m-branding, it is essential that the identity of the brand matches the unique aspects of the mobile phone, such as being a forerunner, technologically advanced, or communicative. For example, the Coca-Cola Company used the mobile channel in a promotion campaign in Finland in order to make the promotion more appealing to the young target group and to communicate the link between the company and continuous innovation.

In general, a thorough identity mapping of the brand (e.g., Aaker, 1996) needs to be performed in order to understand whether the unique aspects of the mobile phone link with the brand associations that the particular brand strives for. Without the link, results of m-branding campaigns are likely to be disappointing.

Brand Assets and the Mobile Channel

Aaker and Joachimsthaler's (2000) framework of brand assets can also be applied to m-branding. In particular, three of the assets can be targeted directly for improvement with the m-branding methods that are available today. These are brand awareness, brand associations, and brand loyalty. Perceived quality, in our opinion, can be improved only indirectly through brand awareness and brand associations. Consumers are known to use intrinsic cues, for example, taste, shape, and material and extrinsic cues, for example, price, brand name, and store name, to infer product quality (Rao & Monroe, 1989). Consumers are unlikely to infer product quality from the company's use of a mobile communication channel, particularly considering the limitations of current displays and the limited possibilities of communicating quality aspects through this channel. However, perceived product quality can be enhanced indirectly by increasing consumers' brand awareness and by creating more positive brand associations.

What the mobile phone offers is an anytime anywhere interactive channel that can be used as a supplement to other communication channels. It can improve brand associations by providing highly relevant information and by utilizing the unique aspects of the mobile phone. All branding activities have to be designed and performed consistently to achieve improvements to the brand assets. Marketers also need to be realistic and understand that some assets are easier to develop than others; brand loyalty is the most difficult. We will now turn the attention to the specific m-branding methods that are available for brand managers and marketers today and show how they can affect brand assets. Concrete examples are provided.

M-Branding Methods

Based on the literature (e.g., Keskinen, 2001; Newell & Newell Lemon, 2001; Pura, 2003), industry sources (The Forrester Report, 2001; Ovum Limited, 2000, 2001), one of the authors' own work experiences, and discussions with other industry experts, four currently existing m-branding methods were identified. These are (1) sponsored content, (2) mobile CRM, (3) different forms of mobile advertising, and (4) a mobile portal.

Sponsored Content

Ovum Limited (2001) states that "all formats of wireless marketing message can be delivered either standalone, or as an accompaniment to content" (p. 112). Sponsored content refers to content that is sponsored by an established and/or well-recognized brand and which can be requested by the end-user via a mobile phone. According to Aaker and Joachimsthaler (2000), sponsorship creates exposure for the brand and develops the consumers' brand associations. Sponsored content affects brand awareness, or recall, through brand exposure and brand associations by associating positive characteristics with the brand, for example, innovativeness, advancement, and mobility. In addition, consumer behavior, though not necessarily long-term loyalty, is initiated when consumers act on the advertisement. Based on one of the author's personal observations and experiences of m-branding campaigns, consumers appear to be more receptive to advertisements that are presented together with content, compared to when being exposed to stand-alone wireless advertisements. According to Ovum Limited (2001), sponsored content is also perceived as less intrusive. Marketers should use sponsored content when they have a well-defined target market (e.g., football enthusiasts) and when they want to differentiate the company as an innovative sponsor.

A concrete example of sponsored content could be as follows. At any given time, spectators watching a hockey game at home could request the interim results of other hockey games through a specific Short Code Number (SCN, a special phone number used to identify a text message based mobile phone service, e.g., 17817) displayed on the TV screen. The SCN could be offered by the sponsor of the hockey game, for example, a sports equipment retailer. The message that the consumer receives could include a special offer or discount on certain items at the store together with the requested content (results of the other hockey games). In this example, the consumer pulls or requests the content via Short Messaging Service (SMS, a text messaging service). The advertising message that accompanies the content can be ignored or acted on.

Mobile CRM (M-CRM)

Mobile CRM means using the mobile channel for customer relationship marketing or management. A key to any brand building effort is gaining knowledge of the relationship that exists between the customer and the brand. Customers have gained more power to choose how they wish to interact with companies; therefore, new tools for customer communication are demanded (Newell & Newell Lemon, 2001). Customers want to be treated individually and receive personalized value. Marketers need to develop alternative ways of reaching customers with timely, relevant, and highly personal information of high value, and they also need to provide customers with a means for timely feedback (Newell & Newell Lemon, 2001). The mobile phone can be used for this purpose. If m-CRM is to have an effect on brand loyalty, per se, marketers need to engage in a dialog with customers and personalize their brand communication messages. For example, the Belgian home improvement store, Brico, collects vital information about customers by inviting "its Discount Club members to join a continued SMS dialogue" (The Forrester Report, 2001, p. 11). Those who sign in receive questions on home improvement, and by answering the questions, they receive discount coupons or invitations to equipment demonstrations. An eventual goal of m-CRM is to empower customers and let them be in charge of the brand communication and interaction (Newell & Newell Lemon, 2001).

Marketers should consider using m-CRM when a highly personalized and timely dialog with customers is desired and when they want to differentiate the customer service or product offering with an innovative method of interaction. Marketers should be cautious, however, since a long-lasting dialog with customers is challenging to obtain. It will most likely engage only those customers who are highly involved in the product or company. Furthermore, marketers should pay close attention to consumer privacy issues and to handling consumers' concerns regarding how their personal information is used (Phelps, Novak & Ferrel, 2000).

Mobile Advertising

Advertising positively affects customers' purchase frequency, their behavioral loyalty toward the brand (Jones, 1998), but it also affects the brand awareness and brand associations (Aaker, 1996; Aaker & Joachimsthaler, 2000; Keller, 1998). However, companies need to recognize the fact that the mobile phone is perceived as inherently personal and that advertising needs to be based only on consumers' explicit permission to receive commercial messages. For example, in Finland, the law prohibits all but opt-in mobile marketing.

The content of advertisements can be independent of time and place or tailored to a specific context. An advantage of the mobile phone is that it is possible to develop interactive advertisements that cannot be used in traditional media. A disadvantage is that traditional goals of advertising, such as creating emotional connections with customers, cannot be easily transferred to the new medium (Newell & Newell Lemon, 2001). However, as mobile devices continue to develop, the possibilities for communicating the emotional dimensions of the brand will improve. In general, marketers should use mobile advertising to support traditional advertising media. The mobile phone is a highly targeting but also supportive channel, which can break through the clutter created by other media and which can engage consumers into immediate action.

Based on the literature (e.g., Keskinen, 2001; Ovum Limited, 2001), mobile advertising can be divided into four different forms: (1) content, (2) transaction, (3) feedback, and (4) location-based advertising.

Content-Based Advertising

Content-based advertising means that the advertisement itself provides content and value to the consumers (Keskinen, 2001), not that the ad is accompanied by content. An example would be an advertisement for a restaurant that includes its menus or a cinema advertisement that includes a detailed listing of films, show times, and film reviews.

Transaction-Based Advertising

In transaction-based advertising, the advertisement is a promotional offer made by the marketer (Keskinen, 2001), and consumers are encouraged to respond directly to the advertisement by buying the product or service or requesting more information. The format of these advertisements can be *push* or *pull* (Keskinen, 2001). In the *push* format, the advertisement is sent to consumers' end-devices based on permission. In the *pull* format, consumers request the product/service by sending an SMS to a number that has been advertised in another media, for example, on the Internet, TV, or at the point-of-sale.

To be effective for various services or products, transaction-based advertising needs to be combined with mobile payment solutions, giving consumers the possibility to react immediately by paying for the product with mobile payment methods. Currently, mobile payment is underdeveloped, but according to a Finnish report by the Ministry of Traffic and Communication (2003), it is expected to increase rapidly from 2004 onward. The report presents statistics by Gartner Research, which estimates that North America will lag behind Western Europe and Japan in mobile payment at least until the year 2007. The value of mobile payment is expected to rise from less than $5 billion in 2003 to reach $30 billion in both Europe and North America by 2007.

Feedback-Based Advertising

In feedback-based advertising, consumers identify themselves by sending an SMS to a specified SCN, thereby informing the marketer about their interest in the product or service (Keskinen, 2001). It differs from transaction-based advertising in that consumers do not make purchases directly through the mobile phone. According to The Forrester Report (2001), it could be a one-off pull campaign, inviting consumers to participate via SMS in a promotion by marketing it on product labels or in radio advertisements. As a practical example, The Forrester Report (2001) mentions Frito-Lay, which prints codes inside its snack bags and encourages buyers to participate by sending the given code, via SMS, to a SCN in order to win a prize. According to The Forrester Report (2001) this way "Frito-Lay acquires a mobile phone number with the receipt of each SMS, allowing it to build its own database of mobile users for future push campaigns" (p. 8). Consumers declare an interest in the brand and permit the company to engage in dialog with them.

Feedback-based advertising is perhaps the most commonly used mobile advertising method. To give another example, the Coca-Cola Company used feedback-based advertising in its Red Collection campaign in 2001. In the campaign, Finnish consumers were encouraged to buy Coca-Cola, to register for the promotion and collect codes that could be redeemed for points. The points could then be exchanged for branded rewards (e.g., CD holders, sun glasses, bags, MP3 players). The more points the consumer earned, the better promotional gifts they could obtain. The possibility of participating via the mobile phone was communicated through various channels, such as the Internet, product labels, billboards, TV, and radio. The case description by Pura (2003) of an advertising campaign for a Finnish chocolate bar is another example of feedback-based advertising.

Location-Based Advertising

Location-based advertising is still in its infancy. This is due, among other things, to the inefficiencies of networks and, in many countries, a lack of clear laws on when and how consumers can be targeted with marketing messages related to their location. Therefore, location-based advertising is currently of minor relevance for branding, but it may become more important in the future when technology improves, clear guidelines are established, and the cost of positioning decreases. Although location-based services exist, there is a lack of practical examples of location-based advertising. One example is a location-based service offered by the Swedish-Finnish telecommunications operator TeliaSonera. Together with a Finnish ski resort, Levi, TeliaSonera offers their customers the possibility of opting in (by opt-in consumers give their permission to be positioned) for a location-based service package, the LevInfo-service, that provides travelers at the ski resort with information on services and service providers in the ski resort area. The location-based service package includes information and advertisements on restaurants, bars, evening activities, and weather reports. Based on their location, consumers receive three to four news and commercial messages per day. The service ends immediately upon leaving the Levi area (or after four days of the initial service request), but it can also be terminated at the ski resort by sending an SMS to a specified SCN provided by the service provider. The service is marketed on the Web, TeliaSonera's own mobile portal, and various point-of-sale materials at the resort.

Although location based advertising sounds attractive, there are challenges that need to be surpassed when large-scale services are offered. According to Ovum Limited (2001):

The data sharing [costs associated with positioning consumers] and massive processing power this would require (regularly comparing the locations of hundreds of thousands, or even millions of users with hundreds or thousands of outlets, and combining it with profiled or data privacy information) have to be weighted against the ease of substitutes— such as having someone handing out leaflets. (p. 109)

Furthermore, marketers have to be careful not to send too many messages per day or week, in order to avoid "spamming" or exceeding the consumer's tolerance level due to constantly beeping mobile phone receiving new text messages as the consumer changes locations.

Mobile Portal

A mobile portal is defined by Ovum Limited (2000) as a "wireless portal [that] concentrates users around communications, content and other applications that are accessible over a wireless network using a handheld terminal" (p. 9). The mobile portal can also be described as a Web site that can be accessed through both fixed line (i.e., the Internet) and wireless (i.e., mobile devices) connections and which offers consumers a variety of services, mobile content, and applications to handheld devices, such as mobile phones and Personal Digital Assistants (PDAs). According to Aaker and Joachimsthaler (2000), fixed line Web sites can be a key part of brand-building efforts due to their ability to convey important information, communicate experiential associations, influence other brand building activities, and thereby affect different brand dimensions. Mobile portals, however, have not been used for brand building. Though several explanations may be given for this, the foremost is that consumers in general possess unsophisticated mobile devices that cannot fully utilize the interactive content of the mobile portal. Mobile portals have also been described as being ahead of their time. With the spread of more advanced phones, mobile portals may well increase in importance, particularly among the young and active phone users.

Wireless or mobile portals should not be elementary versions of fixed line Web sites. They need to be designed so that the unique aspects of mobility are realized, such as (1) mobile communication, (2) expediency–anytime anywhere, (3) real-time interactivity, and (4) independence of location (Ovum Limited, 2001). Aaker and Joachimsthaler (2000) offer guidelines for brand-building Web sites that can be applied also to mobile portals. A brand-building Web site should (1) generate positive experiences for the consumer, that is, deliver value and be easy to use, interactive, personalized, and timely; (2) echo and sustain the brand and its identity; (3) act in synergy with all other communication channels; and (4) grant access to the brand's most loyal customers. In other words, mobile portals, like brand-building Web sites, should offer a home for the most loyal customers with strong brand commitment. On the whole, the portal should reflect and support the brand identity, communicate brand associations to the potential and existing customers, and give customer a reason for repeat visits. It needs to be part of an integrated marketing strategy (i.e., wholly integrated with the Web portal, other communication channels and materials, and all marketing campaigns).

Although mobile portals have not been widely used for branding, some advanced examples can be found. One example is the Yahoo! mobile portal (*http://mobile.yahoo.com*), which offers a variety of services and content that the customer is already familiar with from Yahoo!'s fixed line Web portal. Thus, the Web and mobile portal give a seamless appearance and work extremely well together. Consumers are provided with a way to be attached to Yahoo! as a brand and to its services, regardless of time and location. After registering their mobile phone numbers, consumers can, for example, create mobile alerts on different topics (e.g., breaking news, auctions, sports, e-mail, horoscopes, stocks, news, and weather), download games to be played both via the Internet and the mobile phone, and play virtual or fantasy games, download and upload personal photos, use instant messaging (IM), or read and write e-mail in real-time. Although it could be argued that Yahoo!'s mobile portal is just an extension of Yahoo!'s fixed line Web content, it does show how mobile portal can act in synergy with the Web. It sustains the Yahoo! brand and encourages interactivity and personalization. It also provides access for loyal customers independently of time and place.

Managerial Implications

As can be discerned from the previous discussion and examples, m-branding is still in its infancy. M-branding has only, to a minor extent, been used in cross-media advertising to support brand-building activities. Although m-branding can be used to build and support brand assets, the differential effect of the mobile channel is difficult to determine.

Because consumers perceive their mobile phones as inherently personal, it is important for companies not to clutter it with unwanted messages that will drive the customer away from the brand. M-branding of any form should always be based on an opt-in decision by customers. The brand identity needs to be considered when making a decision on whether to use the mobile channel or not. Managers need to analyze whether customers perceive it as a positive experience to interact with the company and its brands through their mobile phone. Identity mapping (compare Aaker, 1996, for Brand Identity System), for example, can be used to evaluate which parts of the identity the mobile phone corresponds with and whether the target market fits the use of the channel. When the brand personality is characterized as young, timeless, and constantly on the move, the mobile phone would be a good fit. In addition, attention needs to be paid to how the mobile phone links to other communication channels to form an integrated communication message. The mobile phone cannot work as a stand-alone in branding. Without careful considerations of all aspects of the campaign, m-branding will be only a trial-and-error strategy with little chance of succeeding.

Since mobile marketing is a relatively new phenomenon, it catches consumers' interest and has the ability to effectively increase brand awareness and create positive associations to a relatively unknown brand. However, it is important for managers to be aware of the limitations as well as the opportunities of mobile marketing. For example, the interactivity that is required from consumers forms a threshold for participation. Only consumers who feel that it is worth their while will react, and feedback-based advertising campaigns might attract prize seekers rather than interested loyal customers. Furthermore, since m-branding is a new phenomenon, the technique is not always equal to the promises, and the technical failure of campaigns cannot be ruled out.

Although brand loyalty and engagement are generally strived for, it is difficult to discern these results from easily available campaign statistics. Advertising primarily creates awareness and associations, rather than brand loyalty, per se. Mobile portals and m-CRM support loyalty by offering valuable services to the customer, perhaps services that cannot be obtained by other means. To engage in m-CRM, marketers need a sufficiently accurate database of consumer information, including mobile phone numbers and permissions by consumers, to perform mobile marketing. M-CRM is a relationship supporting service that needs to be tailored to the customers' individual needs; they should feel delighted when they get a brand-related message on their mobile phone.

Table 1. M-branding objectives and supportive media for m-branding methods

M-branding	Brand Assets	Supportive Media
Sponsored Content	1. Brand Awareness (recall) 2. Brand Associations 3. Brand Loyalty	Depending on the sponsor and content, e.g., TV and Radio, Indoor/Outdoor Billboards, Indoor/Outdoor Events, Product Labels, Point of Sale, Print Media, Direct Marketing (DM), and the Web.
M-CRM	1. Brand Loyalty 2. Brand Associations	DM (e-mail and ordinary mail), the Web, Mobile Portal, and Personal Selling.
Mobile Advertising	1. Brand Associations 2. Brand Awareness (recall, and for Transaction-Based Advertising brand recall-boosted brand recognition) 3. Brand Loyalty	Content-based advertising needs support from highly visible Printed Media, Billboards, and the Web. Transaction-based advertising needs support from easily available printed instructions. Feedback-based advertising should be well integrated with all the Media in the campaign. Location-based advertising is best supported by Local Billboards and Media, Company Web sites, and Tourist Information.
Mobile Portal	1. Brand Loyalty 2. Brand Associations 3. Brand Awareness (recall)	The Mobile Portal should be advertised in all the Media used by the company, same as the Company Web site.

The costs of a campaign (direct costs vs. costs incurred from possible failure) should also be considered. Depending on the methods used and the comprehensiveness of the m-branding campaign, the incurred costs are competitive in relation to more traditional branding methods (i.e., mobile marketing vs. ordinary mail marketing) and, in many cases, lower. Thus, m-branding provides a cost efficient way of supplementing other branding activities. Ideally, as marketers get more experience of m-branding, their mindsets will be altered so that the mobile channel is incorporated as a permanent part of the company's brand building efforts. Table 1 provides a summary of the brand asset objectives for each m-branding method and of media that might be used in combination with m-branding.

Future Trends in M-Branding

The worldwide growth of GSM network coverage, upgrades to GPRS network, the building of third generation (3G) networks, and the development of more advanced mobile services have, taken together, triggered a discussion of the imminent convergence of the Internet and mobile telecommunications technologies and their potential commercial use (Han, Ahn & Skudlark, 2002; Keen & Mackintosh, 2001; Nilsson, Nuldén & Olsson, 2001). In the future, when mobile telecommunication and the wired Internet converge, it will offer marketers new opportunities for utilizing mobile devices in branding.

As mobile devices become more advanced offering better interactivity features, higher data processing power, faster connections, and better access to the Internet, marketers will be able to design campaigns that incorporate voice, moving pictures with color, and interactive features, such as games and instant messaging. With better user experiences, it will be easier to catch consumers' attention and involve them in the campaign. However, as more companies start to use the mobile channel, the novelty value for consumers will wear off, and it will become increasingly difficult to engage them in a brand dialog. Hence, companies need to improve their ability to segment customers and to design different branding objectives for different target audiences. Consumers who are already committed to the brand may need another type of communicative strategy than the variety seekers or those with a divided loyalty to several brands.

Further research on m-branding and its advantages and limitations from both a managerial and consumer perspective is clearly needed. Interesting research questions would be, for example, (1) Are there consumer segments that are particularly receptive to m-branding and, if so, what characterizes them? (2) Are consumers' emotive reactions toward a brand affected by m-branding methods? (3) To what extent can different m-branding methods achieve different brand awareness objectives (i.e., recall and recognition), either as a stand-alone media or as an accompaniment to more traditional media? (4) Is m-branding equally effective for low- and high-involvement products?

Conclusions

This chapter has discussed m-branding as an emerging phenomenon in brand management and has provided an overview of the m-branding techniques that are available today. M-branding is assumed to affect brand equity through brand assets. However, the currently available techniques have a limited effect on all brand assets and primarily affect brand associations. Considering the current shortcomings of both techniques and phones, marketing managers need to beware of having too high expectations regarding m-branding effects. Reasons for being cautious include: (1) technology failures and the high penetration of low-end handsets, (2) the unfriendly user interface of most mobile phones today, (3) companies' reluctance to do a thorough brand identity analysis to assess the suitability of m-branding, (4) unfamiliarity with different m-branding techniques and limited knowledge of how the mobile phone relates to other channels of communication, (5) marketing professionals' use of the wired Internet as a benchmark for evaluating the effectiveness of m-branding techniques, although the two media are completely different, and finally (6) consumers' privacy needs and how they are being met in branding campaigns need to be considered.

The Internet as a marketing medium has received much attention in both business and research communities, whereas little attention has been paid to the development of marketing through mobile devices. This is an omission that needs to be addressed by further research.

References

Aaker, D.A. (1991). *Managing brand equity: Capitalizing on the value of a brand name*. New York: Free Press.

Aaker, D.A. (1996). *Building strong brands*. New York: Free Press.

Aaker, D.A., & Joachimsthaler, E. (2000). *Brand leadership*. New York: Free Press.

Balasubramanian, S., Peterson, R.A., & Jarvenpaa, S.L. (2002). Exploring the implications of m-commerce for markets and marketing. *Journal of the Academy of Marketing Science*, *30*(4), 348-361.

Barnes, S.J., & Corbitt, B. (2003). Mobile banking: Concept and potential. *International Journal of Mobile Communications*, *1*(3), 273-288.

Crockett, R.O., & Reinhardt, A. (2003, December 22). America zooms in on camera phones. Suddenly the market is hot – and Asian companies are grabbing it first. *Business Week,* European Edition, 30-31.

The Forrester Report. (2001, December). *The marketer's guide to SMS*. Amsterdam, Netherlands: M. de Lussanet.

Han, S.P., Ahn, J.H., & Skudlark, A. Convergence phenomenon and service-network matrix. Retrieved October 25, 2004, from *http:// www.its2002.or.kr/pdffiles/papers/197-hansangpil.pdf*

Heinonen, K. (2004). Reconceptualizing customer perceived value – The value of time and place. *Managing Service Quality, 14*(2/3), 205-215.

Jones, J.P. (1998). *How advertising works: The role of research.* London: Sage.

Kapferer, J.-M. (1997). *Strategic brand management: Creating and sustaining brand equity long term* (2nd ed.). London: Kogan Page.

Keen, P.G.W., & Mackintosh, R. (2001). *The freedom economy: Gaining the mCommerce edge in the era of the wireless Internet.* Berkeley, CA: Osborne McGraw-Hill.

Keller, K.L. (1998). *Strategic brand management: Building, measuring and managing brand equity.* New Jersey: Prentice Hall.

Keskinen, T. (2001). *Mobiilimarkkinoinnin käsikirja* [Mobilemarketing handbook]. Helsinki, Finland: The Association of Finnish Advertisers.

Lindstrom, M. (2001). *Clicks, bricks & brands.* London: Kogan Page.

Ministry of Traffic and Communication. (2003). Mobiili lähimaksaminen – nykykäyttö ja tulevaisuus. [Mobile payment – current usage and the future] (Report No. 22/2003). Helsinki, Finland.

Newell, F., & Newell Lemon, K. (2001). *Wireless rules: New marketing strategies for customer relationship management anytime, anywhere.* New York: McGraw-Hill.

Nilsson, A., Nuldén, U., & Olsson, D. (2001). Mobile media: The convergence of media and mobile communications. *Convergence The Journal of Research into New Media Technologies, 7,* 34-38. Retrieved October 25, 2004, from *http://www.viktoria.se/nulden/publ/pdf/ convergence.pdf*

Nordman, J., & Liljander, V. (2004). MSQ-Model. An exploratory study of the determinants of mobile service quality. In S. Krishnamurthy (Ed.), *Contemporary research in e-marketing, vol. 1* (pp. 93-129). Hershey, PA: Idea Group Publishing.

Ovum Limited. (2000). Wireless portals: Business models and market strategies. Zoller, E., pp. 1-252. Retreived 2001, from *http://www.ovum.com*

Ovum Limited. (2001, June). Wireless marketing: Rhetoric, reality and revenues. Nelson, R.; Ward-Dutton, N.; Brash, C.; & Giddings, E., pp. 1-314. Retrieved 2001, from *http://www.ovum.com*

Phelps, J., Novak, G., & Ferrel, E. (2000). Privacy concerns and consumer willingness to provide personal information. *Journal of Public Policy & Marketing, 19*(1), 27-41.

Pura, M. (2003). Case study: The role of mobile advertising in building a brand. In B.E. Mennecke & T.J. Strader (Eds.), *Mobile commerce: Technology, theory, and applications.* London: Idea Group.

Rao, A.R., & Monroe, K.B. (1989, August). The effect of price, brand name, and store name on buyers' perceptions of product quality: An integrative review. *Journal of Marketing Research, XXVI,* 351-357.

Reichheld, F.F. (1996). *The loyalty effect: The hidden force behind growth, profits, and lasting value.* Boston: Harvard Business School Press.

Repo, P., Hyvönen, K., Panzar, M., & Timonen, P. (2004, January 5-8). *Users inventing ways to enjoy new mobile services – the case of watching mobile videos.* Proceedings of the 37th Hawaii International Conference on Systems Sciences, Big Island, Hawaii.

Storbacka, K., Strandvik, T., & Gronroos, C. (1994). Managing customer relationships for profit: The dynamics of relationship quality. *International Journal of Service Industry Management, 5*(5), 21-38.

Chapter XI

Geographic Information Systems (GIS) in E-Marketing

Mark R. Leipnik, Sam Houston State University, USA

Sanjay S. Mehta, Sam Houston State University, USA

Abstract

The primary purpose of this chapter is to introduce the reader to Geographic Information Systems (GIS) based technologies and applications within the broad domain of E-marketing. Numerous e-marketing examples are presented from diverse industries that will assist the reader in better understanding the various uses and applications of GIS technologies over the Internet. While the use of GIS technologies in e-marketing is in its infancy, it is hoped that the compilation of information through personal interviews, research articles, and personal visits to the various Web sites will assist in validation of the technology within the marketing paradigm.

Introduction

Within the broader context of utilizing the Internet to market products, conducting marketing research, and engaging in other e-marketing applications and/or activities, there is a subset of applications involving the technology of Geographic Information Systems (GIS). This chapter introduces the reader to the structure and functions of GIS, sources of software and data for GIS, and the availability and characteristics of GIS for use by e-marketers. Finally, we will examine specific examples of cutting edge applications of GIS and the Internet in several industries including: tourism, real estate, market research, business-to-business e-commerce, and online provision of geospatial data and services. The objective of this chapter is to familiarize e-marketers with GIS technologies and provide them with another tool for strategic decision making.

Geographic Information Systems

While GIS technologies have been in existence for approximately 40 years, its adoption by marketers and diffusion within marketing is a relatively newer phenomenon (Tomlinson, Calkins & Marble, 1976). Historically, the most common applications (and original uses) of this technology have been in the areas of natural resources management, infrastructure and facilities management, and land records management. More recently, public utilities and municipal governments have embraced the technology (Goodchild, 1991). Today, GIS is being applied to literally thousands of disparate applications from mapping crime incident locations to tracking nuclear submarines (GIS World, 1997). While the technology itself is complex, marketers can use and apply the technology with relative ease once they understand the general structure, functions, and sources of both data and software that are readily available.

Structure

GIS are a suite of computer programs designed to store, analyze, manipulate, and create output like maps, charts, and reports from geographic data linked to descriptive attribute data (Burrough, 1986). The geographic data are typically features that can be portrayed on maps or in aerial photography and stored in a special topological structure. Within this unique format, coordinates, scales, projections, and geometric interrelationships are explicitly maintained. Descriptive attribute data is stored in a tabular format and linked to the geographic data. For every geographic feature, such as a point representing a customer or unit location, a line representing a street or a polygon representing the boundaries of a census tract, there is a corresponding record in a database table.

GIS data is structured into a series of layers. These layers are co-registered (i.e., have the same spatial extent) coordinate system and projection. For example, a GIS might have a layer of zip code boundaries, a layer of streets and highways, and a layer of streams, lakes and wetlands (hydrography), a layer of state and national boundaries, a layer of unit (store) locations, a layer of competitor unit locations, and a layer of customer locations. For each distinct layer, there would be (at least one) database table containing as many records as there were features in the corresponding geographic (map) layer to which they were linked. Each record there would be a primary key with a unique numerical identifier that differentiates that record and links the record to the feature (such as a point representing the residence of a customer). There would also presumably be a series of other fields containing applicable information, such as customer social security number (or other unique identifier), customer address, customer phone, fax, and e-mail information, customer credit card account balance, responses to a customer survey, prior sales to that customer, credit report data, and so forth (Samet, 1989).

Function

The complex and somewhat unwieldy topological data structure described above has its compensation in that many unique forms of spatial analysis can be performed on the data stored in the GIS. Thus, one can readily generate buffer zones around features and then select the corresponding records for the features falling within a designated buffer zone around some locations. For example, one could create a one-, five-, and ten-mile buffer zone (i.e., primary, secondary, and fringe trade areas) around every unit of a franchise, like Burger King (Mehta, Leipnik & Maniam, 1999a). Then, all census blocks or block groups falling within these concentric circles would be extracted from the 7,017,427 blocks and/or 229,192 block groups (from the 2000 census) already stored in the Census Bureau's Topologically Integrated Geographic Encoding and Referencing (TIGER) data set (Leipnik, Mehta & Maniam, 2003). Demographic information, such as population, race, age, and gender by block and other data, such as median household income by block group for all persons responding to the census and residing within one, five, and ten miles of every Burger King in the U.S., could then be readily determined.

Another example of a spatial analysis method feasible with GIS would be to determine the areas within five, ten, and twenty minutes travel time from every Ann Taylor Store in the U.S. and Canada. The street data used in this analysis would need to be correctly connected together and would need speed limit data linked to every segment. Once the travel time zones for each store were determined, these could be compared to actual customer locations (if previously geocoded), zip codes within which customers resided, census demographics for the blocks, block groups, census tracts, and so forth falling within these travel time limits and potential competitor locations.

These examples only scratch the surface of spatial analysis methods employing off-the-shelf GIS software and readily available data sets. With respect to the emphasis of this book on e-marketing applications, it is important to note that the spatial analysis methods discussed above can all be conducted interactively (in real-time) over the Internet, Intranet, and Extranet. Many, if not all, of the data sets mentioned above are available for access or purchase over the Internet (Landis, 1993; Peng & Tsou, 2004). Due to both complexity and relative stability of the software, the general relational database structure of GIS software has not changed (significantly) for several decades. Vendors have recently released an object-oriented database structure (Gordillo & Balaguer, 1998).

Sources of Software and Data

The TIGER data set (*http://tiger.census.gov*) is a core component of marketing use of GIS in the U.S. and is available over the Internet on a county by county basis from the Environmental Systems Research Institute (ESRI) (*http:/*

/www.ESRI.com), on CDs in a GIS readable format from Geolytics (http://www.geolytics.com), and interactively it can be viewed and maps generated from it using the Census Bureau's interactive mapping Web site (http://TIGER.census.gov). Also, enhanced TIGER data is available from GDT (http://www.GDT.com). Other useful geospatial data includes street data from Teleatlas (http://www.teleatlas.com) and Navtech (http://www.navtech.com), economic data from Dun and Bradstreet (http://www.DandB.com), and reaggregated and customized data from Claritas (http://www.claritas.com). This geospatial data by itself is useless since a highly sophisticated suite of computer programs (i.e., GIS) is needed to store, manipulate, and analyze it. Today, the leading GIS software is the ARCGIS product line from ESRI. Of particular significance is the ARC Internet Map Server (IMS), a powerful and scalable product designed to serve up maps and other forms of geospatial data (e.g., aerial photography) interactively with a range of spatial analysis tools over the Internet (Spee, P.C., October 5, 2003).

A major competitor of ESRI is MapInfo (http://www.mapinfo.com). MapInfo is very affordable (approximately $250 per copy) and has been dominant in what is referred to as *desktop mapping* applications. Their MapXtreme product is designed to serve maps over the Internet. MapInfo has a strong position in market research firms who typically use stand-alone versions of the software to perform simple mapping applications (e.g., sales volume by territories). MapInfo has recently diversified its portfolio of businesses by purchase of Thompson Associates, a leading market research firm in the franchise site selection and cannibalization analysis area (Mehta et. al., 1999a). MapInfo also has a strong presence in Australia, where the Australian Bureau of Statistics (their census authority) uses MapInfo for its online mapping (Leipnik, Mehta & Seidel, 2002). In Canada, MapInfo has acquired Canmap, the leading purveyor of the nationwide street data (J. Hobson, P.C., October 5, 2003).

As the popularity of GIS in business, in general, and e-commerce, specifically, has expanded, other firms have tentatively entered into the field; these include companies, such as AutoDesk which has dominated the Computer Aided Design (CAD) field and Intergraph which has specialized in high-end turn-key graphics solutions for government and industry for decades. A more significant recent entrant in the field is Microsoft with the Mappoint GIS. Mappoint is fully compatible with the rest of the Microsoft Office suite. At approximately $200 per copy, Mappoint is intended to compete directly with MapInfo. Like MapInfo, it has a rather limited range of functions and a commensurately lower price than ARCGIS (which retails for approximately $1,500 per copy). As of now, Mappoint has a relatively small market share of the approximately $2 billion per year GIS market (GIS World, 2002). Mappoint has a Web services component designed to serve street map data to users around the world. At present, Mappoint has licensing agreements that allow it to serve street data for use in mapping, research, and navigation for 19 North American and European countries. Since Microsoft has expertise in development of Web-based applications (e.g., Explorer, Outlook) and mass marketing application software (e.g., MS Office suites), we might expect that within a few years only two competing visions for Web-based GIS are likely to be dominant: Microsoft's low-end desktop mapping approach designed for the casual user of GIS and ESRI's sophisticated object-oriented ARCGIS and ARCIMS products designed for building geospatial data and performing advanced spatial analysis.

GIS and Marketing

Since its introduction is the 1950s, the 4P (i.e., *place, price, product,* and *promotion*) framework has become the dominant paradigm within marketing. Most authors of marketing textbooks within the disparate discipline (e.g., *Principles of Marketing, Sports Marketing, Internet Marketing, Global Marketing,* and *Marketing Management*) use it as the primary structure. Therefore, the marketing mix variables (4Ps) will be used to validate the significance of GIS within e-marketing.

While some researchers have argued that the old adage, "three most important things in retailing are location, location, location" is still valid, others have argued that the advent of the Internet has tended to reduce the importance of one of the Ps, place (Wang, Sui & Lai, 2004). They contend that it is now possible to outsource many activities (e.g., call centers, technology support, and Web development) to remote locations in countries like India. It is also possible to buy products from e-retailers located in other countries (e.g., *http://www.Amozon.com.uk*), thereby questioning traditional theories and laws (e.g., Reilly's Law of Retail Gravitation, Huff's Model).

However strong or weak the above stated argument, the power and importance of *place* should not be underestimated (Sui, 2002). GIS can be used to analyze the demographic, economic, and infrastructure related to a given place. These factors often determine the appropriateness of locating a new retail unit or making a real estate investment. For example, the number of persons residing within a specified distance or travel time would help to determine the profit potential of a trade area. The presence of employers, hotels, airports, highways, arterial streets, public transportation, inter-modal links, and so forth would help determine the potential customer base for drive-through or walk-up franchise systems. Physiographic factors (e.g., proximity to flood zones or earthquake faults) can also have an affect on the long-term value of an investment in a particular area.

The fact that GIS allows analysis of issues regarding place should be of no surprise to anyone, but envisioning how GIS can be used in relations to the other three Ps is not so obvious. It is probably easiest to find instances where *promotion* is taking advantage of spatial analysis afforded by GIS. For example, GIS data on traffic counts is used to set billboard and truck-side advertising rates and select billboards for placement of specific advertisements (Leipnik, Mehta & Newbold, 2003). GIS data on the demographics of network and cable television and radio markets combined with viewership data are also used to set advertising rates. Customer addresses and unit locations along with zip code boundary data can be effectively used in targeting direct mail advertising campaigns such as that of Ace Hardware Stores (Harder, 1998).

GIS can also be used to determine the optimal *product* mix to offer in a given retail outlet. This has particular relevance for supermarkets and supercenters (e.g., Wal-Mart, Fiesta) that carry an excess of 50,000 items. Local variations in ethnicity and socioeconomic status can have a tremendous impact of what products (e.g., baby foods) will sell profitably. GIS has also been used to study the placement of products within stores in relation to traffic, shelf space, and setting slotting allowance rates.

GIS can also be used to set or at least influence the *prices* charged for products. Geographic pricing is commonly used, where price varies due to distribution costs. For example, FedEx sets delivery charges via zones, the boundaries of which are based on GIS analysis. The same concept has pertinence in business-to-business (B2B) marketing, where products that are not universally produced and/or are heavy and cumbersome to deliver (e.g., automobiles, coal, steel, electricity) have set charges that vary from region to region. It is obvious that pricing structure can be made more complex by other factors including delivery bottlenecks, variable demand, level of competition, and local, state, and national laws and regulations.

GIS and E-Marketing

While there are numerous disparate definitions of e-marketing within the academic literature, the best one in our opinion is the one that utilizes and expands the official AMA definition of marketing. E-marketing is "the use of electronic data and application for the planning and executing the conception, distribution, promotion, and pricing of ideas, goods, and services to create exchanges that satisfy individual and organizational goals" (Strauss & Frost, 2001, p. 454). Most e-marketing strategies used today are designed to attract, develop, and enhance mutually beneficial relationships with customers and suppliers by enhancing value and creating satisfaction and loyalty (Mehta, Dalal & Maniam, 2001). While there are numerous applications and uses of GIS in e-marketing, we identified a few leading industries: e-tourism, B2B e-commerce, e-real estate industry, e-banking, and so forth.

E-Tourism

One particularly fruitful area for the utilization of interactive GIS-based mapping over the Internet is in the field of tourism. The precedent for use of the Internet in travel is well established with a large volume of airline, hotel, rental car companies, and other travel-related reservations being made online. Since travel by its very nature has a geographic component/association (e.g., selection of lodgings in an unfamiliar city), it is obvious that the use of maps is valuable (if not essential). With the advent and evolution of the Internet, there is demand to access these maps online. Some examples of Web-based GIS in the field of e-tourism include imbedded GIS within Web sites like Expedia, Travelocity, Goto, Mapquest, Alta Vista, and so forth.

The Web sites of several national tourist offices contain interactive mapping capabilities, including that of the British Travel Authority (*http://www.visitbritain.org*). The Visit Britain Web site features an initial screen that asks users to identify the country they are coming from. It then links to a page in that country's official language and features travel consultants and toll-free numbers relevant for that country. The gateway page features a prominent map of Great Britain, that when selected, invokes a GIS-based interactive mapping application. Clicking on an area within the map opens a pop-up with a larger scale map of that region portraying major roads and towns. Clicking on a town then zooms to local roads, lakes, and rivers and allows users to click on icons

for lodging, attractions, and transportation. For each attraction, lodging establishment, and so forth there is a link to a descriptive page which assesses the
attribute data linked to each feature. This site utilizes a generalized version of
the British Ordinance Survey Master Map GIS, which are arguably the best
national GIS for a major country (The Swiss and Finnish GIS are comparable).

Another very popular feature of the Visit Britain Web site is the interactive
Movie Map. This feature uses some of the same framework and features as the
Interactive Map. However, it features maps with icons and linked imagery
showing the locations where scenes in popular British films have been shot
(Shaw, P.C., October 19, 2003). Outside the U.K., tourism related information is being disseminated using interactive mapping in Israel, Australia, and
New Zealand. In Canada, arguably the country where GIS was invented in the
early 1960s (Tomlinson et al., 1976), the seventh largest industry (tourism) is
using GIS actively including pioneering wireless Internet applications using
handheld PDA-based approaches for travelers in major cities like Toronto.

Web-based GIS technology is not exclusively for the wealthy nations. The
State of Goa (India) has been used as a test market for Tourism GIS over the
Web. In order to more easily disseminate information about access to tourist
attractions, travel kiosks with touch screens and Internet access to a server with
interactive maps are being established in a number of hotels in tourist locations,
such as Agra (Taj Mahal), New Delhi, and Bombay. In Malaysia, Creative
Advanced Technology Corporation has developed a Web portal termed *http:/
/www.virtualmalysia.com*. This Web site features interactive GIS-based
maps along with satellite imagery, draped over terrain models. In Zimbabwe,
the Tourism Authority has used ArcView Network Analyst to determine the
best routes to attractions from major towns and hotels and has posted such
maps on its Web site.

B2B E-Commerce

One of the leading providers of geodemographic data is Claritas (*http://
www.Claritas.com*). Claritas specializes in providing geodemographic data
such as reaggregated census statistics as well as providing specialized and
proprietary geospatial data to market researchers and others interested in the
analysis of the distribution of potential customers or units. Claritas is perhaps
best known for taking the responses from the decennial census and reaggregat-
ing them, either by census subdivisions with new attributes or by zip code zones.
The value of reaggregation is that zip codes are often more readily available

from consumers (than street address data) and are far less likely to contain erroneous information. Instead of having to characterize a census tract or block group as containing hundreds of variables, it is possible to combine selected variables that have a high correlation with consumer buyer behavior. It is then possible to classify the block group, census tract, city, county, Metropolitan Statistical Area (MSA), and so forth as containing a given proportion of consumers that have certain characteristics (for segmentation).

The most memorable feature of Claritas' PRIZM and their Microvision products is the catchy phrases used to describe each demographic segment. At present, there are approximately 60 groups in PRIZM, ranging from *Blueblood Estates* (White or Asian race, median annual household income of $158,000, refined tastes such as yachting and reading the New York Times, and purchasers of luxury automobiles), through groups like *Pools and Patios, Bohemian Mix, Golden Ponds, Shotguns and Pick-ups* all the way down to *Hardscrabble Farms* (White or Native American race, $22,000 median annual household income, interest in reading *Guns and Ammo*, watching Country Music TV, and purchasers of recapped tires).

Besides data derived from the census, Claritas offers a huge range of other geospatial data, all of it available to market researchers over the Internet through their Web portal (Moore, P.C., October 19 2003). Recently, Claritas has inaugurated a feature of its Web site intended to foster the use of interactive GIS in the area of B2B e-marketing. This site features access to a Business Facts data set on characteristics, locations, and contact information (including Web site addresses) for business by MSA. The data set also includes the North American Industrial Classification System (NAICS), employment levels, initial public offerings, data from annual 10-K reports and various other SEC filings, consumer spending data by zip codes, shopping center locations and attributes, traffic volume data, and capability to use road data with network connectivity to perform drive time analysis online. Claritas also offers an e-connect service designed to foster B2B e-commerce.

E-Real Estate

SSR Reality is a Real Estate Investment Trust (REIT) specializing in catering to the intermediate-term investment needs of large institutional customers. It must service a highly sophisticated clientele that typically makes investment decisions concerning allocation of portfolios. SSR currently manages over $6 billion in assets for such institutional investors as pension plans and insurance companies. In order to serve its existing clients and win over new investors, SSR must differentiate itself from a wide range of investment firms. It does this through superior analysis of real estate investment opportunities in the sixty-two markets in which it invests. Its analysis is heavily dependent on spatial data to generate predictions of return on investment for commercial and multifamily residential properties. SSR uses GIS, regression, and geostatistical modeling to conduct analysis of these properties.

SSR utilizes Web-based applications in three ways. The first use is to access geospatial data (e.g., digital aerial photography from publicly available sources such as Terraserver *http://www.terraserver-usa.microsoft.com*) for use on its server. Second, it serves associates in the field with a vast quantity of data from its secure server usually for making presentations to clients or for field assessments. Third, SSR has developed an application termed Real estate On Line (ROL) that allows clients to view potential properties and perform limited spatial analysis online.

Several aspects of SSR's use of geospatial technologies place this firm on the forefront of market research application of GIS. For example, instead of relying on universally available data sets (e.g., TIGER), SSR utilizes a wide range of data (e.g., Dun and Bradstreet data, internal data on rents, proprietary data on occupation rates, and rents for competing properties). SSR does not simply generate maps based on summary statistics, instead multivariate statistical analysis and two-dimensional spatially weighted Kriging models (Anselin, 1998) are employed to generate probabilistic estimates with a spatial component that can in turn be mapped and distributed over the Internet. Thus, instead of mapping median income by census tract as a predictor of market potential, SSR can generate an estimate of annualized return-on-investment adjusted for predicted inflation based on a series of weighted factors, such as rents and occupancy rates in similar properties, depreciation adjusted for tax effects, employment in surrounding businesses, and life cycle stage for a given area where there is a property. This approach requires a far higher level of sophistication and a far greater investment in resources (i.e., to gather and/or purchase the data from which these predictions are ultimately derived). However, the proof of the efficacy of this approach is the increasing inflow of funds into this REIT and the emphasis that SSR places on geostatistical and Web-based approaches in its own corporate strategy and in presentations to clients (Scherer, P.C., November 11, 2003).

E-Banking

RPM consulting (*http://rpmconsulting.com*) specializes in sophisticated analysis of the spatial distribution of customers, unit locations, and infrastructure variables. Clients include Western Exterminator, Dallas Federal Teachers Credit Union, South Coast Federal Credit Union, and so forth. Banking marketing is a good example where fostering of an ongoing relationship based on mutual trust and a real understanding of customer needs is critical (Mehta, Maniam & Mehta, 1999). Spatial factors also play a role in the establishment of such relationships. People typically patronize businesses that are easy to access. Traditionally, that access has been based on physical proximity. Thus, one would patronize a local bank in the town or city in which one worked or lived.

The advent of branch banking, Automatic Teller Machines (ATM), and Internet banking has altered this paradigm. In the Los Angeles area, many employees of major defense contractors (e.g., Lockhead-Martin) formed the South Coast Credit Union (SCCU). It was both specific to an industry (aerospace) and a geographically defined region (i.e., South Coast area of Los Angeles/Orange counties) where the defense contractors were concentrated. While proximity is relevant, recent technological advances allow financial institutions to practice relationship management with depositors, even when they move out of close proximity to a branch.

In the case of SCCU, six urban branches serve customers. The depositors are usually well educated, have high incomes, are technologically savvy, and somewhat mobile. In recent years, aerospace firms have relocated facilities to the Antelope Valley and Inland Empire Regions of Southern California. The customers of SCCU have frequently taken jobs at these facilities and moved to be closer to their jobs. In traditional banking, once a customer moves and no longer patronizes a branch, the relationship devolves to the impersonal form. The opportunity to practice relationship with these distant customers diminishes. SCCU decided to use GIS to practice Customer Relationship Management (CRM).

For each branch, a trade area was established linking that branch with those customers that are closer to it than to any other branch (S. Lackow, P.C., October 22, 2003). This procedure had the effect of frequently reassigning customers from the branch in which they may have opened an account to the nearest branch, as well as creating territories for each branch in which they can pursue new business opportunities while not causing undue dislocation to customers. Looking at this data spatially, a bank official might make several decisions. One would be to locate new branches to better serve these far-flung customers. Another would be to encourage Internet-based banking, use of ATM's, and smart Kiosks at these more remote locations. Use of the Internet to reach and maintain relationships with relocated customers is an opportunity highlighted by this type of mapping. Thus, a customer in Palmdale (in the Antelope Valley) might receive customized e-mails informing him of the proximity to the nearest branch where cost-free ATM transactions can be made and offer Internet only promotions (e.g., repossessed cars for sale in the customers area).

Web-Based Geocoding and Spatial Data Mining

ESRI is not only the world leader in GIS, but it is positioning itself to be a leader in Web-based services for marketing and business decision making. Specifically, what ESRI is offering are a variety of fee-based and free services that allow small- and medium-sized businesses to access the latest geospatial data and the most sophisticated tools and expertise related to issues like geocoding (Mehta, Leipnik & Maniam, 1999b). Instead of struggling to procure geospatial data, clients can access the latest enhanced census data or up-to-date street centerline data without the trouble of purchasing voluminous and costly data sets. One can submit a simple ASCII file, spreadsheet, or database table with customer names, sales totals, and zip code data. ESRI will then use Zip Code Tabulation Area (ZCTA) data combined with TIGER data to provide a map and attribute database showing the zip code boundaries that enclose every customer and the median income, race, population, and other demographic characteristics for that area. This is an attractive option, since the cost of current versions of data sets can run into the thousands of dollars (lct alone the difficulty in error checking and geocoding the zip code data). The resulting map and report can be billed on a per item basis that for large data sets comes out to a few cents for each customer geocoded and mapped. Even more expensive data sets (e.g., Teleatlas data) can be used to geocode customer addresses that can then be matched against the census block groups that the customer resides within.

It is hoped that the above stated examples can provide e-marketing with another tool to both better understand and appreciate GIS technologies and their Web-based applications. Unfortunately, just like other tools used for marketing decision making, GIS too has numerous limitation.

Limitations

There are several significant limitations to the use of GIS over the Web. These include the ongoing problem of the accuracy of spatial data, perhaps best exemplified by the inaccuracy of many of the streets used as the basic framework of the TIGER data (the cornerstone of much geodemographic analysis). If data in the U.S. is inaccurate, data in Canada and Europe is often prohibitively expensive. The situation in many other parts of the world is an absence of geospatial data necessary in marketing. Another limitation of this field is the potential for concerns about privacy restricting availability of data down to the most detailed levels (e.g., blocks, individual households). Finally, the travails of Enron provide a cautionary example of the limits of the use of Web-based and geospatial technologies. Enron was a major user of GIS as well as a key player in Internet-based marketing of energy services (and eventually of broadband capacity itself). Enron's sophisticated use of GIS, optimization, financial analysis, game theory, and so forth, frequently involving fraudulent or conspiratorial and predatory energy trades over the Internet, only temporarily boosted profits at the expense of long-term stability and the viability of the entire energy trading industry.

Future Research

The application of GIS over the Internet in marketing is so new a concept that little research has as yet been done on it. However, recently several books have appeared discussing the technology. These include (1) *Internet GIS: Distributed GIS for the Internet and Wireless Networks* (Peng & Tsou, 2004); (2) *Serving Maps on the Internet* (Harder, 2002); and (3) *Connecting Our World: GIS Web Services* (Tang & Selwood, 2003). These books, academic journal articles, and personal interviews with practitioners strongly indicate that there are several areas of fruitful research.

For example, there are numerous spatial analysis methods available. Which of these methods are most appropriate for use over the Internet and how valid are the predictions yielded by each method in comparison to the complexity of the algorithms and the type and quantity of data required? Another example is how best to convey information in relatively small maps. That is, what are the appropriate cartographic design considerations, and how can they be incorporated into Web sites to offer users easy to create and easy to comprehend cartographic products, legends, scale bars, titles, and source information? This is even more an issue when the screen becomes smaller (as in the evolving area of m-commerce).

Personal communication systems (e.g., cell phones, PDA), equipped with Global Positioning Systems (GPS) technologies and Internet-based GIS, have many potential applications in the future. For example, a customer in downtown New York City can identify the closest Starbucks outlet, place an order, pay for it with his credit card, and have the coffee ready to be picked up (or delivered). Future applications of CRM include the creation of virtual groups among spatially concentrated customers. For example, destination retailers (e.g., Outlet Malls, Bass Pro Shops, L.L. Bean) who maintain a database of their regular customers, could alert all customers (via e-mail) within a certain zip code (through a map) that Mrs. Jane Doe plans to patronize the destination store next weekend, and she is willing to give customers a ride to the store.

Conclusions

Powerful and complex technologies (e.g., GIS and WWW) have in the past been largely separate and independent. Recently, synergies between the technologies have begun to be exploited by e-marketers. For example, GIS software has been adapted to serve maps on the Internet. GIS data is being accessed from the Internet. Multiple sources of data are being submitted and retrieved from GIS programs, over the Web.

As these and other technologies converge, numerous additional applications will be developed. Many of the most exciting examples of the convergence of GIS, GPS, Internet, and so forth will involve e-marketing applications. Ultimately, it will be the ability of interactively generated maps, delivered over a medium (e.g., PDAs) with a user-friendly interface that will drive demand, uses, and appreciation of GIS in the future.

References

Anselin, L. (1998). GIS research infrastructure for spatial analysis of real estate markets. *Journal of Housing Research, 9*(1), 113-125.

Burrough, P.A. (1986). *Principles of geographical information systems.* Oxford, UK: Oxford Science.

GIS World. (1997). *GIS World sourcebook.* Fort Collins, CO: GIS World.

Goodchild, M.F. (1991). Geographic information systems. *Journal of Retailing, 67*(1), 3-15.

Gordillo, S., & Balaguer, F. (1998). Refining an object-oriented GIS design model: Topologies and field data. *Proceedings of the Sixth ACM International Symposium on Advances in Geographic Information Systems,* 76-81.

Harder, C. (1998). *GIS means business.* Redlands, CA: ESRI Press.

Harder, C. (2002). *Serving maps on the Internet.* Redlands, CA: ESRI Press.

Landis, J. (1993). *Profiting from geographic information systems.* Ft. Collins, CO: GIS World Press.

Leipnik, M.R., Mehta, S.S., & Seidel, R. (2002, December 10-13). Geodemographic data: An international perspective. *Proceedings of the 2nd International Conference on Electronic Business (ICEB).* Taipei, Taiwan.

Leipnik, M.R., Mehta, S.S., & Maniam, B. (2003, March 12-14). *An introduction to TIGER 2000.* Proceedings of the Marketing Management Association Spring Educators Conference, Chicago, Illinois.

Leipnik, M.R., Mehta, S.S., & Newbold, J.J. (2003, June 9-13). *Truck side advertising: An appraisal.* Proceedings of the European Applied Business Research Conference, Venice Italy.

Mehta, S.S., Leipnik, M.R., & Maniam, B. (1999a, March 6-7). Using geographic information systems in franchising. Proceedings of the 13th Annual Society of Franchising Conference, Miami Beach, Florida.

Mehta, S.S., Leipnik, M.R., & Maniam, B. (1999b). Application of GIS in small and medium enterprises. *Journal of Business and Entrepreneurship, 11*(2), 77-88.

Mehta, S.S., Maniam, B., & Mehta, S.C. (1999). Relationship banking: A multinational bank's application of relationship marketing. *Journal of Business Strategies, 16*(2), 121-134.

Mehta, S.S., Dalal, G., & Maniam, B. (2001). Customer relationship management strategies for the Internet. *Academy of Information and Management Sciences Journal, 4*(1), 19-33.

Peng, Z., & Tsou, M. (2004). *Internet GIS: Distributed geographic information systems for the Internet and wireless networks.* New York: Wiley.

Samet, H. (1989). *Design of spatial data structures.* New York: Addison-Wesley.

Strauss, J., & Frost, R. (2001). *E-marketing* (2nd ed.). Upper Saddle River, NJ: Prentice Hall.

Sui, D. & D.J. Rejski (2002). The Emerging Digital Economy. *Environmental Management, 29*(2), 155-163.

Tang, W., & Selwood, J. (2003). *Connecting our world: GIS Web services.* Redlands, CA: ESRI Press.

Tomlinson, R.F., Calkins, H.W., & Marble, D.F. (1976). *Computer handling of geographic data.* Geneva, Switzerland: UNESCO Publications.

Wang, J., Sui, D.Z., & Lai, F.Y. (2004). Mapping the Internet using GIS: The death of distance hypothesis revisited. *Geographical Systems*, forthcoming.

Section IV:

E-Marketing Legal Challenges

Chapter XII

Legal Online Marketing Issues:
Opportunities and Challenges

Michael T. Zugelder, Old Dominion University, USA

Abstract

This chapter will review several emerging issues associated with e-marketing from the perspective of Web sites including intellectual property, information management, and contracting. E-marketers in the years to come should not only consider the growing e-marketing opportunities but also be cognizant of the growing body of law governing marketing online. While the future may be bright and uniformity and fairness may prevail, current domestic and global law facing e-marketers is varied and the potential basis for liability. The law must be considered wherever e-marketers direct their efforts.

Introduction

The Internet now opens the world to e-marketers. No longer are international marketing and trade the monopoly of the giant multinational corporations. With the Web's low barriers of entry and its incredible ability to distribute information and facilitate contacts to nearly all points on earth, all marketers can ply the art and science of e-marketing on a global basis to develop and enjoy both foreign and domestic markets. Consumer choices in e-marketing abound in developed countries well-connected to the Internet. In the U.S. alone, retail e-marketing has reached 38 million consumers, accounted for $95 billion of sales in 2003, and is forecasted to reach $230 billion by 2008 (Forrester Research, 2003). E-marketing will also make those choices profoundly available in developing countries that, prior to e-marketing efforts, did not enjoy the reach now made possible.

Interest in promoting online marketing and commerce has also caused lawmakers worldwide to consider how law governing contract, intellectual property, tort, and consumer protection can be changed to facilitate the productive use of the new medium. The goal of many of these efforts is law that is more transparent, uniform, and commercially tolerant, yet ethically responsible.

The other side of the story is that, globally, law governing these important legal issues remains significantly dissimilar so that while e-marketers large and small can enjoy the new worldwide opportunities of the Internet, they also face parallel challenges presented by a myriad of existing laws that govern e-marketing. This chapter will review several important legal issues that arise from e-marketing and the use of marketing Web sites.

Intellectual Property Concerns

E-marketers using the Internet and Web sites are operating on a mass medium which can infringe on intellectual property rights of others. Due to the global character of the Internet, infringements can occur worldwide, causing terrific exposure to e-marketers. At the same time, great benefits can be obtained by e-marketers successfully utilizing intellectual property law to protect their trademarks, copyrights, trade secrets, and patents.

Trademark

A trademark is any symbol used in commerce that connects the product to its source in the mind of the consumer. It is marketing's unique intellectual property, providing legal protection for branding and all forms of advertising slogans and logos. Use of another's trademark without permission can cause trademark infringement. Indeed, trademark infringement has been the leading cause for litigation and arbitration on the Web. Cases have largely been caused by unauthorized registration and use of another's established trademark as a domain name, either to extort a settlement and sale of the domain name to the trademark owner, which has been addressed by anti-cybersquatting statutes, or through suits for trademark dilution where the unauthorized domain name use has caused a potential for confusion or disparagement of the trademark through negative association with the product or service offered by the e-marketer. So, in *Hasbro, Inc. v. Internet Entertainment Group, Ltd.* (1996), a U.S. court enjoined the use of the domain name Candyland.com by a pornographic Internet Web site because it diluted Hasbro's Candyland trademark in its board game. Intentional adoption of a domain name for purposes of extorting a settlement has been addressed by anti-cybersquatting statutes like the U.S. provision adopted in 1999. The Anti-Cybersquatting Consumer Protection Act of 1999 addresses the problem for individuals as well as trademark holders by allowing the court to take jurisdiction of the suit where the domain name was issued and grant an order canceling or transferring the domain name to the plaintiff. A similar approach, utilizing Internet arbitration rather than court litigation, has been provided by the Internet Corporation for Assigned Names and Numbers (ICANN).

E-marketers should also steer clear of inappropriate Web page techniques used to drive business at the expense of trademark holders. These techniques have included: (1) deep linking, (2) metatags, and (3) framing. Deep linking–hyperlinks to the interior of another company's Web site which bypass the home page and much of its banner advertising–has caused litigation. In *Ticketmaster Corp. v. Microsoft Corp.* (1997), Ticketmaster sued and forced settlement by Microsoft, which had provided hyperlinks into the interior of Ticketmaster's Web site falsely suggesting an affiliation between the two companies and confusing consumers as to the origin of information. Metatags–the invisible code embedded in HTML used to create Web sites and assist search engines–have triggered litigation against e-marketers for using another's trademark within the metatags. Thus, in *Playboy Enterprises, Inc. v. Calvin Designer Label* (1997), the Court enjoined use of plaintiff's trademark in its domain names, metatags, and home page text, announcing such abuses constituted trademark infringement. Framing–the technique of placing Web sites of another within the Web site of the e-marketer–has likewise brought allegations of trademark infringement, as demonstrated in *Washington Post Co. v. Total News, Inc.* (1997).

On the other hand, e-marketers are encouraged to utilize trademark protection to the extent provided by law. In the U.S. and elsewhere, e-marketers are entitled to register their entire Web site as a *trade dress*, where the association of color, design, shape, or graphics of a Web site are sufficiently associated with the product or service. Trademarks should be registered and domain names reserved. Trademarks, in fact, make very good domain names, unifying existing marks and marketing efforts with the mission of e-marketing.

Copyright

Copyright provides the owners of creative works with the exclusive right to exploit them. It does so by forbidding the unauthorized copying, distribution, and derivation of the work and provides a basis for civil suit and, in some instances, criminal prosecution. Today, copyright protection begins at the instant that a creative work is placed in any fixed medium, such as a CD, DVD, photo, or computer file. Copyright law now extends to a wide variety of creative works relevant to e-marketing, including Web sites, electronic data-bases, and computer software. Copyright is increasingly recognized as a universal right of property. Globally, over 100 countries have signed the Berne Convention, and a number of other treaties and conventions have been signed with the goal of making copyright law more uniform and copyright registration and enforcement less onerous. Contrary to the belief held by some, copyright protection extends to the Internet and e-marketing. Stated simply, merely because a creative work has been displayed on the Internet does not place it in the public domain. Rather, as in the past, copyright law has been modified as technology has changed. In the U.S., the passage of the Digital Millennium Copyright Act (DMCA) and similar legislation worldwide, confirms copyright law applies to the Internet and e-marketing. High-profile U.S. Internet litigation involving the music and movie industries demonstrates that copying of creative works without permission can result in liability. Still, in many instances, copyright's exact role in protecting property placed on the Internet is both uncertain and evolving. Take, for instance, the dismissal of a criminal prosecution in Norway against a teenager who created a DeCss de-encryption program and placed it on the Internet, which allowed users to access and copy DVDs online without permission (Mellgren, 2003). Note also the proliferation of peer-to-peer music file swapping networks, even in the face of hundreds of suits brought by the Record Industry Association of America (RIAA) and record companies holding the copyrights. While reported suits like *UMG Recording, Inc. v. MP3.com, Inc.* (2000) result in large judgments, and cases like *A&M Records v. Diamond Multi-Media Systems (Napster)* (2000) have found online music file sharing services vicariously liable for copyright infringement, recording executives are now fearing peer-to-peer networks as significant competitors to the point of hiring online companies, such as Beverly Hills-based BigChampagne, to track popular peer-to-peer networks to determine the popularity of specific artists and titles (Veiga, 2003).

Such results are not shocking given copyright's evolution in light of mass media technology. As in other countries, U.S. copyright law has been amended to take into account technology-driven changes and create a fair balance between society's interest in making free information available and the copyright owner's exclusive rights. For example, the DCMA creates a new basis to sue those who would defeat encryption protection, while at the same time providing Safe Harbor suit immunity for Internet service providers drawn into court for copyright infringements caused only by their customers' postings.

Trade Secrets

Trade secrets require no registration, yet provide indeterminate protection to any secret that allows competitive business advantage. Trade secrets include anything confidential, from customer lists and marketing strategies all the way to the secret formula for Coca-Cola and KFC's eleven herbs and spices. E-marketers, those selling high technology in particular, use trade secrets in lieu of patent because of low cost and short life cycle of many Internet e-marketing techniques and technologies. The other side of the equation is that trade secrets can lose their protection by public disclosure, and nothing discloses like the Internet. Once a trade secret is on the Internet, it is usually lost and saved only by immediate and forceful litigation. For that reason, marketers like Ford Motor, Raytheon, and chip maker Cadence Corporation have sued to enjoin unauthorized posting of their trade secrets by disgruntled employees, competitors, and others (Cundiff, 2002).

Patents

A patent protects inventions and other applications that are sufficiently novel by providing the right to exclude others from making, using, or selling a patented item. Patents are widely recognized worldwide, but their award and enforcement remain largely nation-based. Patent law has become extraordinarily important to e-marketers in the U.S. due to the recognition of the business method patent, which protects unique ways of conducting e-marketing. Since the decision in *State Street Bank v. Signature Financial Group* (1999), which first confirmed this type of patent protection for financial software applications, e-marketers have applied for and received business method patents and brought infringement actions to protect them. In *Amazon.com v. Barnes & Noble.com* (1999), Amazon brought an injunction against Barnes & Noble to prevent them from using the Amazon One-Click business method. Though the case was ultimately settled, others have licensed the One-Click business method from Amazon. Aside from this contract-making technique, business method patents have been granted for a variety of e-marketer applications, such as *Priceline.com*'s online reverse auction; Cybergold's consumer reward system, Attention Brokerage model, allowing consumers to review Internet ads; Netcentives' frequent flyer mile exchange reward system; DoubleClick's DAR model for information collection; and many other applications.

E-Marketing Information Management

Like brick-and-mortar marketing, a crucial function of e-marketing is information management. Because e-marketers function on a global mass medium, they must consider the many laws that regulate, control, and sometimes forbid the collection or dissemination of information. These laws address civil injury or tort caused by consumer misrepresentation, privacy violation, defamation/disparagement of person, company, or product, trespass, and intentional damage to property. Such laws provide a right of civil action for damages and court orders and may also be the basis for criminal prosecution. Because both torts and crimes involve serious injuries that go to the heart of a country's role in protecting its citizens, jurisdiction over e-marketers is usually triggered. Further, because they are nation-based, parochial, and usually not the matter of treaty, they are widely divergent in their terms. For all these reasons, they pose a significant concern to e-marketers and must be considered.

E-Advertising Practices

Once online, e-marketers must be aware of the potential application of advertising and trade laws various countries will apply to their goods and services on the Web. Violations can occur simply from providing information. Contract making and/or product distribution is not required. Because an e-marketing advertising campaign is frequently global, it will be virtually impossible to check for legality against all of the consumer protection laws nations currently enforce. Still, there is some clear guidance for e-marketers to consider. Obviously, fraudulent, deceptive, and unsubstantiated advertising claims, false endorsements, and the like violate both self-regulatory codes as well as most consumer protection laws. In the U.S., the Federal Trade Commission (FTC) has for nearly a decade taken a very aggressive position toward misrepresentation online. Consumer protection rights have also been addressed by other agencies in the U.S., including the Securities & Exchange Commission, regulating misrepresentations in connection with electronic public offerings and trading, and the Federal Drug Administration (FDA), in conjunction with U.S. Customs, monitoring, regulating, and prosecuting misrepresentation by e-marketers of pharmaceutical products.

Complaints to consumer protection agencies are expected to mount as e-marketing grows. According to a survey by the National Association of Consumer Agency Administrators, representing government agencies that protect consumers in the U.S., complaints concerning the Internet, e-commerce, and e-marketing are now ranked seventh on the top ten lists for consumer complaints–tied with complaints against cable and satellite companies and telecommunications businesses (Salant, 2003). Consumer complaints about e-marketing include misrepresentation of goods and delivery terms, outright fraud on auction sites such as e-Bay, questionable ISP billing practices, unsolicited e-mail from mortgage lenders, and work-at-home get-rich-quick schemes (Anfuso, 2003).

False and deceptive advertising by e-marketers has caused the FTC to bring over 182 cases since 1994, stopping consumer injury from e-marketing schemes and misrepresentation at an estimated annual savings of nearly a quarter-million dollars a year. Recent actions have included suits for false rebates, deceptive claims concerning free and low cost computer systems, misrepresentations regarding delivery dates and contests, and failure to adequately disclose long-distance telephone charges and limited Internet access to complain (Cendali & Ellen, 2002).

E-marketers, and particularly those in the U.S., are cautioned to consider the FTC's Internet advertising guidelines, which generally require all advertising claims to be truthful, not misleading, properly substantiated, and fair both in their terms and in their conspicuous clarity. Specifics include the appropriate layout of hyperlinks, font sizes, colors, disclosure boxes, and so forth, that can be downloaded. Global compliance with such regulations is even more daunting, but practical steps can be considered to reduce risk of civil action or criminal prosecution, including avoidance of heavily regulated products, such as alcohol and tobacco, and avoidance of certain types of advertising targets that have posed great concern for lawmakers, including health, children, and financial advertising where advertising campaigns are concentrated.

E-Marketing Privacy Concerns

Modern marketing requires the collection of information concerning consumers' interests. E-marketers have become very effective information collectors, and technologies such as cookies have created an Internet subindustry. The valid interests of marketers to collect information, however, must be balanced against a growing concern from consumer interests and their governments about what information is being collected and how it is being disseminated. In the U.S., e-marketing privacy is largely the subject of FTC regulation, together with myriad federal and state privacy statutes. Broad statutes protecting very sensitive information concerning children, finance, and health matters have recently been passed. The Children's Online Privacy Protection Act (COPPA) limits collection of information from children under the age of 13 and requires parental consent.

The Gramm-Leach-Bliley Act forbids disclosure of certain financial information collected by e-marketers and others and, in other instances, requires a consumer's right to opt out, preventing information dissemination. The regulations under the Health Insurance Portability Accountability Act (HIPAA) are aimed at the U.S. health industry and forbid transfer of medical information for purposes other than treating and billing with the intent to protect sensitive medical information from being used by e-marketers and others. Further, the FTC, as the major U.S. governmental enforcer of consumer privacy, has aggressively prosecuted numerous actions against e-marketers for unfair information practices arising from violations caused by information collection contrary to posted Web site privacy policies. For instance, an action against GeoCities was brought for the collection of information concerning children, even though the Web site stated no such information would be collected or shared without permission. Arising from the GeoCities action, the FTC articulated its current four fair-information principles, including (1) notice–e-marketers must disclose their information practices before collection—encouraging a published policy regarding collection and use; (2) choice–consumers must be given the option on how information is collected with at least the option to opt out of such collection; (3) access–consumers must be able to review and contest the accuracy and completeness of data collected; and (4) security–data collectors must take reasonable steps to ensure information collection for consumers is accurate and secure from unauthorized disclosure.

Globally, several comprehensive privacy initiatives have been established. The European Union's Directive on Privacy Protection is one of the most significant efforts in this regard. Effective in 1998, the directive requires E.U. member states to adopt legislation to protect privacy as it relates to personal data and forbids its collection and dissemination unless specific restrictive requirements are met, including, in many instances, the consent of the consumer. Such opt in consent requirements are significantly more stringent than U.S. privacy laws. As a result, the U.S. has had to reach a Safe Harbor agreement with the E.U. in order for U.S. e-marketers to collect information. Compliance with the Safe Harbor agreement has meant that e-marketers must maintain a privacy policy and follow it. Many e-marketers have elected to contract with third-party seals in order to comply with such requirements.

E-marketers must also understand that suit exposure shall continue to exist (even where a privacy policy is followed) for information collected without explicit consent through technology, such as cookies. In the U.S., several class actions against the largest cookie merchant, DoubleClick, including *In Re DoubleClick Privacy Litigation* (2001), have resulted in significant financial settlements, and privacy groups are still pressing for legislation to ban all involuntary data collection.

Finally and obviously, e-marketers should refrain from intentional collection through clearly illegal means, such as hacking and electronic eavesdropping, which support not only civil action, but also criminal prosecution.

Defamation and Disparagement: Cyber Libel

An untrue statement of fact damaging to reputation which is published may be considered defamation to a person or disparagement to a business, product, or service. E-marketers will frequently be considered publishers, and defamation will be considered online libel and judged in much the same way as marketers using any other mass media. Cyber libel presents a vast basis for exposure and has led to suit in foreign countries. For instance, in *Dow Jones & Co. v. Gutnick* (2002), the High Court of Australia permitted a libel suit against Dow Jones to proceed in Victoria, based on its posting of a *Barron's* magazine article online. The Court concluded jurisdiction was appropriate, applying the place of the injury rule. In another highly publicized action, French antidefamation groups brought suit against San Diego-based Yahoo.com in France for violation of French group defamation law arising from advertisements posted and transactions completed on Yahoo!'s auction site by collectors of Nazi memorabilia. The French court, in *Yahoo.com, Inc. v. LaLigue Contre Le Rascime et Anti-Semitisme* (2000), ruled that Yahoo!'s accessibility in France established jurisdiction there, and Yahoo! was liable for violation of French civil group defamation law. The French court directed Yahoo! to filter its site and assessed damages on a per-day basis for its failure to do so. As the French plaintiffs proceeded to collect the judgment in California, Yahoo! brought suit and obtained a U.S. court decision, affirmed on appeal, holding that Yahoo! and its users' First Amendment rights would be violated if the French judgment were enforced in the U.S. The *Yahoo* case demonstrates the potential for global exposure wherever speech in the form of promotion, advertising, or otherwise is placed on the Internet. The case, however, is only one of many that will be brought in the future and far from settles the risk posed by defamation laws around the world that take a far greater pro-plaintiff approach to defamation. One such law, the U.K.'s Defamation Act of 1996, has caused many customers and e-marketers alike to bring libel suits in the U.K. McDonald's, as an example, obtained a significant result against Web site libelers in England. Per the statute, English courts are also not hesitant to find jurisdiction.

E-marketers should consider the national laws of countries targeted by advertising and promotion campaigns and steer clear of violations. Take, for example, comparative advertising, long encouraged by the FTC as a useful source of information for consumers. Product comparison sites are now one of the fastest growth categories for e-marketing. Sites like RealTime, NexTag, PriceGrabber, BizRate, and others offer consumers side-by-side product and price comparisons, together with consumer reviews, and are exceedingly popular (Liedtke, 2003). However, comparative advertising might be considered trade disparagement elsewhere.

Likewise, e-marketers should be knowledgeable of Internet legislation passed to protect them from libel suits. In the U.S., ISPs are immune from liability caused by customers placing defamation on their systems under Section 230 of the Communications Decency Act (CDA). E-marketers, including AOL and Yahoo!, who fit the definition of ISPs, are entitled to rely on this immunity. Such was the case in *Sidney Blumenthal v. Matt Drudge and AOL* (1998). Drudge, the author of the electronic column called The Drudge Report, was sued for defamation by Sidney Blumenthal. Blumenthal also joined AOL as a codefendant. AOL, however, was successful in obtaining dismissal of the suit based upon the CDA. The obvious intent of the statute is to encourage Internet information dissemination, and, to the extent e-marketers fit within its protected class, they should utilize the statute. Other such statutes exist in other countries and should be considered.

E-marketers may also find themselves damaged from competitors' false advertising claims or from consumer complaint sites viciously criticizing their products or services. Complaint sites have numbered in the thousands on the Internet. Outside of the U.S., strong tactics to address false advertising and disparagement should be considered. Within the U.S., e-marketers should understand that the First Amendment protects speech rights far more than elsewhere so that advertising commentary and consumer criticisms are far more protected and harder to reverse.

Trespass Violations

Trespass law provides the owner or legal occupier of property with the exclusive right of possession and the concurrent right to exclude others who enter on property or interfere with its use without permission. Because trespass is an intentional tort and also a crime, the theory of trespass has been an underlying basis for many civil and criminal antihacking statutes in the U.S. and elsewhere. For example, the Computer Fraud and Abuse Act of 1986, a pre-Internet statute, criminalizes intentional physical and electronic trespass of computer systems or data and has been frequently relied on to prosecute Internet hacking. Similarly, the Electronic Communications Privacy Act of 1986 has lately been asserted against e-marketers for the surreptitious placing of cookies.

More recently, legislators are responding to the growing problem of unsolicited electronic bulk advertising, or spam. Marketers are cautioned to consider the growing number of antispam statutes. Law is on the books, for instance, in Australia, Denmark, and Italy requiring e-marketers sending spam to obtain consumer opt in authorization. Congress has also spoken by passing the CAN-SPAM Act of 2003, which allows suits by ISPs against advertisers whose products or services are advertised in e-mail spam. But the legislation has drawn strong criticism for not going far enough to protect businesses and consumers (Sturdevant, 2003). Still, the legality of spam is increasingly in doubt, and e-marketers should govern their actions accordingly.

E-Marketer Contracting

Contracts and product distributions by e-marketers through Web sites are increasingly the paramount goal. Contract formation and performance online remains generally governed by the laws of many nations. Generally in the U.S., the sale of goods is governed by the Uniform Commercial Code (UCC), and the sale of services is governed by case law decisions making up the U.S. common law. Contract law has flexible rules with practical design so that, in many instances, the terms of the contract are left to the parties to form. Such practical rules have been extended by analogy to e-marketing contracts. Thus, the traditional contract rule that a contract can be formed in any reasonable manner, including the conduct of the parties, has been used to support contract acceptance by mouse click (*Caspi v. The Microsoft Network, LLC*, 1999). The established rule that a contract can be accepted and need not be in writing if payment or shipment of goods has been made also serves to support paperless electronic e-contracts. Likewise, the traditional rule that a meeting of the minds must occur with informed intent releases buyers from hidden Web site terms (*Specht v. Netscape Commun. Corp.*, 2002).

Where needed, new contract law has been created. Recently, the Uniform Computer Information Transaction Act (UCITA) has been proposed for adoption. So far, only Virginia and Maryland have done so. UCITA would govern the sale and licensing of information and computer-related products, frequently the object of e-marketing. UCITA specifically makes the familiar "clickwrap" contract for software binding. Other U.S. statutes, such as the Uniform Electronic Transaction Act (UETA), adopted by most states, and the federal Electronic Signature in Global and National Commerce Act (E-SIGN), specifically affirm the validity of contracts made electronically where traditional contract rules might have fallen short.

The principle of freedom of contract generally allows parties to make their own deal. As such, business-to-business e-contracts can and do include a wide variety of innovative provisions which will hold up in court. But when marketing to consumers, the concurrent principles of fairness and informed intent caution e-marketers to carefully consider the design of their Web sites and conspicuous placement of terms provided. Web terms e-marketers intend as binding contract provisions should be clearly and concisely set forth as such. The Web design should encourage consumers to view the contract terms. Related policies governing privacy and any disclosures and disclaimers relating to advertising and other legal compliance should be prominently displayed and carefully thought through. Critically, e-marketers should utilize choice-of-forum and choice-of-law clauses that stipulate the nation or state's law that will govern the contract's interpretation and the court or other forum where a dispute arising from the contract shall be resolved. U.S. decisions generally uphold such clauses so long as they are reasonably related to the e-marketer's principal geographic location and are not unduly burdensome. E-marketers are, however, cautioned that such clauses do not always work to insulate them from jurisdiction. The alternative strategy, at least in the U.S., is to restrict the e-marketing Web site to a passive billboard format, as opposed to an interactive and contract-making one. Further, even in the case of a passive site, choice-of-law-and-forum clauses do not provide a perfect insulation from foreign jurisdiction, such as the E.U., and have typically failed where the consumer has asserted a cybertort, or crime has been committed rather than a dispute based on breach of contract.

Trends and Implications

The following is offered as a summary for consideration by e-marketers. A recap of this discussion is provided in Table 1.

Intellectual Property: Offense and Defense

E-marketers will enjoy the trend toward greater property protection through more uniform laws furthered by treaties and international agencies like the WIPO and WTO. Still, e-marketers must recognize the increased possibility of infringement caused and suffered by e-marketers. E-marketers need to protect

Table 1. Summary of trends/implications and recommended action

Topic	Trend/Implication	Recommended Action
Intellectual Property Concerns	• Increasing protection of IP rights • More infringement suffered through e-marketing	• Register at home and in target markets • Avoid marketing practices known to cause infringement
E-Marketing Information Management/Cybertort Avoidance	• Data collection and disbursement triggers new privacy and antispam laws • Increasing foreign defamation actions	• Abide by home and target privacy and spam laws • Maintain truthful privacy policy, follow safe harbors • Select target market carefully and be aware of defamation suit exposure
E-Marketer Contracting	• Traditional contract rules increasingly recognized • Foreign suit jurisdiction risk by interactive Web site	• Maintain principles of fairness and disclosure • Avoid jurisdiction with passive Web site • Use choice-of-forum clause with interactive Web site

copyrights, trademarks, and patents at home and in each target market by registration. Trade secrets need to be kept off the Internet entirely, and the Internet must be monitored for unauthorized postings. Conversely, e-marketers should avoid specific market-driving practices courts have considered infringement, like deep links, metatags, domain name misuse, and unauthorized file sharing and database exploitation.

Information Management: Cybertort Avoidance

Privacy invasion, trespass, and defamation all arising from information collection or dissemination are increasingly sore subjects that the law around the world has begun to address. Data collection by Web site has triggered stringent E.U. privacy directives and class action and FTC sanctions in the U.S. E-marketers should abide by privacy laws at home and in target markets. Web sites collecting data should post truthful privacy policies and allow, at minimum, the right to opt out of information collection. When marketing to the E.U., compliance with Safe Harbor by third-party privacy registration is recommended. E-marketers must take care in sending unauthorized bulk e-mail, increasingly considered by many countries as a trespass and a fraud. E-marketers should particularly take care in minimizing defamation and product disparagement actions by being selective in target market selection and monitoring the law there carefully.

E-Contracting

The law will continue toward neutrality by increasing application of established tried-and-true contract principles. Contracts with consumers especially require e-marketers to maintain high standards of fairness and full disclosure of contract terms. Foreign suit exposure will depend in large part on the level of Web site interactivity. E-marketers must therefore decide whether to reduce exposure by use of a passive Web site or, instead, enjoy greater market effectiveness with a higher level of Web site interactivity but with the side effect of suit exposure. E-marketers choosing the latter type of Web site are encouraged to employ choice-of-law and choice-of-forum clauses to reduce the cost and likelihood of foreign suits.

Conclusion

This chapter has reviewed several main issues associated with e-marketing and its legal environment. As e-marketing opportunities grow, vigilant attention must remain toward the problems and opportunities associated with intellectual property, information management, and contracting. While efforts are being made to provide a more habitable legal environment with more uniform and transparent law, e-marketers in the near future must also be cognizant that e-marketing is global and that both domestic and global rules can provide a basis for liability.

References

A&M Records v. Diamond Multi-Media Sys. (Napster), 239 F.3d 1004 (9th Cir. 2001).

Amazon.com v. Barnes & Noble.com, 73 F.Supp. 2d 1228 (W.D.Wash. 1999).

Anfuso, D. (2003). Complaints about e-commerce on the rise. *iMediaConnection.com*. Retrieved October 25, 2004, from *http://www.iMediaConnection.com/content/news/120103.asp*

Berne Convention. Retrieved October 25, 2004, from *http://www.wipo.int/clea/docs/en/wo/wo001en.htm*

Caspi v. The Microsoft Network, LLC, 732 A.2d 528 (NJ Sup Ct 1999).

Cendali, D., & Ellen, B. (2002). Advertising and consumer protection on the Internet: How to ensure that your site complies with consumer protection laws. *Computer and Internet Lawyer, 19*(12), 1-24.

Children's Online Privacy Protection Act, 15 U.S.C. § 6501., *et seq.* (2003).

Computer Fraud & Abuse Act of 1986, 18 U.S.C. § 1030 (2003).

Cundiff, V. (2002). Protecting trade secrets from disclosure on the Internet requires diligent practice. *New York State Bar Journal, 74,* 8-16.

Defamation Act of 1996. Retrieved October 25, 2004, from *http://www.hmso.gov.uk/acts/acts1996/1996031.htm*

Dow Jones & Comp., Inc. v. Gutnick (2002). HCA 56 (December 10, 2002).

Electronic Communications Privacy Act of 1986, 18 U.S.C. § 2511, *et seq.* (2003).

Electronic Signature & Global & National Commerce Act (E-SIGN), 15 U.S.C. § 7001, *et seq.* (2003).

European Union's Directive on Data Protection, 95/46/EC. Retrieved October 25, 2004, from *http://www.cdt.org/privacy/eudirective/EU_Directive_.html*

Federal Trade Commission. (2003). Dot Com disclosures: Information about online advertising. Retrieved October 25, 2004, from *http://www.ftc.gov/bcp/conline/pubs/buspubs/dotcom/html*

Federal Trade Commission. (2003). FTC releases survey of identity theft in U.S. Retrieved October 25, 2004, from *http://www.ftc.gov/opa/2003/09/idtheft.htm*

Forrester Research, Inc. (2003, August 5). Forrester research projects U.S. e-commerce to hit nearly $230 billion in 2008. Retrieved October 25, 2004, from *http://www.forrester.com/er/press/release/0,1769,823,00.html*

Gramm-Leach-Bliley Act, 15 U.S.C. § 6801, *et seq.* (2003).

Hasbro, Inc. v. Internet Entertainment Group, Ltd. Case No. C96-139 (W.D.Wash. 1996).

In Re DoubleClick Privacy Litigation, 154 F.Supp. 2d 497 (S.D.N.Y. 2001).

In the Matter of GeoCities, (1998). W.L. 473217.

Liedtke, M. (2003). Reborn shopping.com is product comparison site. *Associated Press.* Retrieved October 25, 2004, from *http://desecretnews.com/dn/view*

Mellgren, D. (2003). Norwegian teenager is acquitted in DVD case. *Associated Press* (January 7, 2003). Retrieved last on December 20, 2004, from *http://www.sfgate.com/cgi-bin/article..cgi?file=/news/archive/2003/01/07/financial0804EST0026.DTL&type=tech*

Playboy Enterprises, Inc. v. Calvin Designer Label, 985 F.Supp. 1220 (N.D.Cal. 1997).

Salant, J. (2003). Car problems top 2003 list of consumer complaints. *Associated Press.* Retrieved last on December 20, 2004, retrieved from *http://www.thebatt.com/news/2003/11/25/News/Automotive.Sales.Top.List.of.Consumer.Complaints-5667724.shtml.*

Sidney Blumenthal v. Matt Drudge & AOL, CA No. 97-1968 (D.C.D.C. 1997).

Specht v. Netscape Communications Corp., 306 F.3d 17 (2d. Cir. 2002).

Standards for Privacy of Individually Identifiable Health Information: Final Rule. 45 C.F.R. Parts 160 and 164 (2002).

State Street Bank v. Signature Financial Group, 149 F.3d 1368, *cert. den.* 119 (S.Ct. 851 1999).

Sturdevant, C. (2003). CAN-SPAM act can't. *eWeek Enterprise News & Reviews*. Retrieved October 25, 2004, from *http://www.eweek.com/article2/0,4149,1238738,00.asp*

Ticketmaster Corp. v. Microsoft Corp., No. 2:97-CV-03055 (C.D.Cal. 1997).

Uniform Commercial Code. Retrieved October 25, 2004, from *http://www.nccusl.org/nccusl/desktopdefault.aspx*

Uniform Computer Information Transaction Act. Retrieved October 25, 2004, from *http://www.ucitaonline.com/*

Uniform Electronic Transaction Act. Retrieved October 25, 2004, from *http://www.nccusl.org/nccusl/desktopdefault.aspx*

UMG Recording, Inc. v. Mp3.com, Inc., 2000 WL 1262568 (S.D.N.Y. 2000).

Veiga, A. (2003). Labels Tap Into Swapping Trends. *Associated Press*. Retrieved last on December 20, 2004, from *http://launch.yahoo.com/read/news.asp?contentID=215540*

Washington Post Co. v. Total News, Inc., Case No. 97 (S.D.N.Y. 1997).

Yahoo.com, Inc. v. LaLigue Contre Le Rascime et L'Antisemitisme, 169 F.Supp. 2d (N.D.Cal. 2001), 198 F.3d 1083 (9th Cir. 2002).

Chapter XIII

Regulatory and Marketing Challenges Between the U.S. and EU for Online Markets

Heiko deB. Wijnholds, Virginia Commonwealth University, USA

Michael W. Little, Virginia Commonwealth University, USA

Abstract

One of the biggest challenges marketers face with e-commerce is the regulatory environment, both at home and abroad. This is especially pertinent for international transactions between the U.S. and the European Union (EU). This chapter attempts to identify and categorize the major global issues involved. It also points out similarities and differences between the U.S. and the EU on such issues as regulation and self-

regulation, taxation, jurisdiction and liability. The purpose of the chapter is to explain the various marketing challenges resulting from specific legal problems. Recommendations are developed for firms contemplating e-marketing to the EU.

Introduction

The Internet's commercial potential for marketers is growing steadily. Forrester Research predicts that by 2004, global online commerce will reach $6.8 trillion for both business-to-business and business-to-consumer transactions (Forrester, 2003). Expanding global opportunities through e-commerce appears especially inviting when traditional channels are too expensive or prohibitive due to numerous foreign country restrictions. This may especially be the case if marketers can reduce costs in their global supply chain, enhance global customer relationships, or build a stronger and more consistent global brand. For U.S. firms of all sizes, the European market is attractive given its increasing interest in and use of e-commerce. With the global online market increasing, the European Union nations represent a substantial market in size and scope. Notably, online users in Germany, France, Netherlands, Finland, and Sweden offer the most promising developments for consumer goods with their high incomes and technology savvy populations. The U.K. and Ireland are especially inviting because they are English speaking nations with strong cultural ties to the U.S. In the future, this market will continue to grow as the EU adds new member countries expanding to a total of 25 member nations.

The Internet is an ideal vehicle to reach the global marketplace that reflects a variety of styles, tastes, and products. Some of the more successful e-commerce firms in the U.S., such as Dell, Amazon.com, Yahoo!, e-Bay, and Land's End, have incorporated Internet technology into their business models. The nature of their business combined with the Internet lends itself readily to global markets. There are, however, new opportunities presented by e-commerce for those willing to adapt their firm's marketing strategy to global e-commerce markets. Increased use of the Internet has enhanced e-marketing practices, such as communicating or selling a large set of products or services to customers. At the same time, despite geographical dispersion, one can inexpensively reach small segments of the population with similar tastes through the Internet. As e-marketers in the U.S. consider foreign markets, they will need to evaluate their current marketing objectives and strategies and adapt them to a particular country or region. These objectives may include:

1. *Attracting new customers:* Web sites provide another channel for e-marketers to promote their goods or services whereby potential customers may browse information that is of particular interest to them. Clearly, foreign customers are more likely to buy a product if it is communicated to them in their own language and caters to local tastes. Those firms that develop local and foreign Web sites may find attendant problems relating to content, language, and technology. Policies should be clearly stated at the Web site to enhance the European customer's experience and build loyalty. For example, policy issues may range from respecting customer privacy by informing them how personal information is used to resolving complaints regarding shipping and returns.

2. *Building customer loyalty of current customers:* The Internet is a communication tool for online marketers to build relationships with current customers. Electronic-customer relationship management (e-CRM) is the management of relationships over the Internet. Studies indicate that different relationship marketing tools (e.g., direct mail, interpersonal communications, and tangible rewards) have different affects on consumer perceptions of a relationship and ultimate loyalty to a firm. Engendering trust becomes a critical factor as firms are required to provide self-disclosure data from customers in the use of discounts, sweepstakes, and other promotions (Fassott, 2003). As we shall see in the following sections, the U.S. and EU have differing approaches as to how such information is collected and how consumers are informed. In this instance, U.S. firms must inform and or gain permission from EU consumers about such practices.

3. *Monitoring competitive trends for local and global markets:* Competition online can come from anywhere because the Internet offers firms greater geographic reach to sell products. Once a competitor is identified, monitoring the online efforts of a competitor's product offerings and pricing provides insight to future strategy. Both the International Trade Administration and Small Business Administration provide information on exporting, country information, and other services (Wilson & Abel, 2002).

4. *Encouraging joint product development and promotions with European partners:* The Internet can encourage interactive communication

with customers and suppliers in developing and promoting new products. There are opportunities here for both B2B and B2C e-marketers. Such arrangements can provide foreign partners who have knowledge and access to their region or country's market (Avlontis & Karayanni, 2000).

There are, however, a number of regulatory issues facing online marketers in the U.S., as they consider opportunities in foreign markets. The purpose of this chapter is to examine such differences regarding regulation and review regulatory policy affecting taxes on Internet goods and services and liability and jurisdictional policies that exist in the EU. Finally, suggestions will be provided for marketing managers in dealing with these issues.

Regulation versus Self-Regulation

The regulation of e-commerce is perhaps the most controversial and unresolved issue that affects the respective positions of the U.S. and the EU on e-commerce trade. Clearly, the U.S. government has, so far, favored as much self-regulation by e-commerce firms as possible. The industry is leading the charge. There are some consumer groups and other critics who have raised some concerns but they have, so far, been overwhelmed by the advocates for self-regulation. The U.S. government has been amenable to allow acceptable international bodies to be created in instances where self-regulation is impractical (e.g., international registration of domain names). The EU and its member states are tilting the other way–towards more regulation, especially to protect consumers. (However, there is a determined effort underway by European e-commerce companies to allow for more self- regulation). This situation is almost the exact opposite of that in the U.S. The circumstances for individual states, both in the U.S. and in the EU will vary to some degree. Unless and until relevant federal U.S. legislation is passed, the situation will remain rather fluid and uncertain. The same applies to the EU where each state is still sovereign with its own legal system, although subject to EU directives when ratified.

A single market with one set of rules for all, open to competition with a minimum of regulation is one of the cornerstones of the EU. In contrast to the U.S., the creation of a single EU market is still a work in progress. The advent of e-commerce has by its very nature created a unique opportunity to accelerate this process. Unfortunately, the initial consumer and business adoption in the EU has generally been slower than that of the U.S. Hence, serious regulatory efforts are off to a late start and have also been hampered by differences between member states. The U.S., however, is experiencing its own discord between the different states and because of its generally pro-deregulation stance has itself been slow to address key regulatory issues (Zugelder et al., 2000).

In comparing the respective positions of the U.S. and the EU on these concerns, it becomes evident that the EU generally favors a more regulated business environment whereas the U.S. prefers as much market discipline and self-regulation as possible. This is especially true for e-commerce, where the U.S. position is to allow this embryonic innovative type of business to flourish in as unfettered an environment as possible. It has been pointed out that this difference is due to fundamentally opposed philosophies; Europeans tend to trust government (more) to keep business in check while Americans tend to favor keeping the government out of business affairs as much as possible. The former is more concerned with consumer protection while the latter believes more in the self-discipline of a free marketplace (Hargreaves, 2000; Waldemeir, 2000).

One example of differing attitudes and practice is the increasing popularity of buying music online from Internet music sites in the U.S., such as Apple Computer, Real Networks, and Best Buy. Of course, a lawsuit by the Recording Industry of America eventually shutdown Napster's free file sharing service in the U.S. and puts continued users at risk. Purchasing online by American consumers from such sites avoids legal hassles and attracting viruses spread by services, such as Kazaa in the Netherlands. More U.S. firms, such as Wal-Mart, are considering entering this market in the U.S. with intentions to offer music at reasonable prices. In Europe, however, such services are severely limited by legal issues. U.S. firms intending to enter this market must pay royalties to collecting societies besides additional taxes from a given country. Purchases of online music by Europeans is further limited to those with credit cards with U.S. billing addresses (Mitchener, 2003).

In conclusion, self-regulation by industry can be effective, but U.S. firms must make visible efforts to comply with serious marketplace problems in Europe, such as data collection, taxation, product liability and jurisdiction issues. Even then it may be necessary for some government intervention by the U.S. and EU.

Regulatory Policy

The concerns referred to above have surfaced in a number of specific regulatory policies which are analyzed below. The positions for each area taken by the U.S. and the EU, respectively, are compared. Table 1 presents areas of direct relevance to online marketers.

Sales Taxation

The taxation of sales is treated differently in the U.S and the EU. In the U.S., a moratorium is still in place on the imposition of (new) and discriminatory taxes on Internet purchases. This does not imply that no sales taxes can be levied on e-commerce transactions. Such taxes may indeed be imposed, just as they may be applied to other direct marketing transactions, such as catalog sales. Most vendors, though, cannot be forced to collect these taxes if they have no permanent establishment or nexus in the purchaser's state. Consumers are supposed to pay the tax on out-of-state sales but this hardly ever happens and collection is rarely if ever enforced. The practical effect is that direct marketers do not collect sales taxes in states where they do not have a physical presence (Schwarz, 2000; Smiegowski, 2000).

Table 1. Specific regulatory positions by U.S. and EU affecting e-commerce trade

Areas	Positions	
	U.S.	EU
Tax - sales	moratorium	VAT - differs between states permanent establishment?
Tax - income	U.S. residence - taxes on global income permanent establishment double taxation treaty with EU	double taxation treaty with U.S.
Liability and replaced jurisdiction control	tentative standard for jurisdiction: degree of interaction uncertainty - courts tend to limit global liability of U.S. firms	consumer home state has business: home-country confusion & contradiction failure of the Hague conference
Marketing Practices	Consumer state regulations?	uncertainty, controversy - see liability above

The EU taxes sales, including those made via the Web and by catalog, through the use of the value-added-tax (VAT) system whereby an item is taxed at each stage of its production or distribution on the value added from the previous stage. Generally, such taxes are included in the final sale price to the consumer without any reference to the tax component. Despite attempts at tax harmonization in the EU, VAT rates differ between the member states (e.g., 22% in France and 15% in Luxembourg). For EU businesses, this is not a major problem since sellers use the rates set by their home state where they (head office) are located. For non-EU marketers, such as U.S. firms, this poses a major challenge. If they have a European presence (e.g., a subsidiary), they too may presumably charge the rates applicable to the subsidiary's home country, such as Amazon.com. Not so for American firms with no such affiliations (e.g., e-businesses exporting directly to consumers). They have to charge VAT based on where the customers live. This means administering different rates for different customer countries—a big challenge, especially for the smaller firms (Barlow, 1999; Utton, 2000).

This problem has now been exacerbated by the EU requirement applicable to all products provided by non-EU e-businesses that are downloaded by European customers over the Internet, such as music, videos, and software as well as Web-based fees. These downloads and services are now also subject to VAT. This affects a variety of American-based online firms including Amazon.com, eBay, and AOL. Again, those U.S. firms without EU affiliations, usually the smaller ones, will be required to apply the VAT rate of the country in which the consumer resides. The EU has taken this step to level the playing field between American and European firms since the former are not subject to any sales taxes in the U.S. while the latter had to charge their EU customers VAT (Grant, 2002; Newman, 2003).

The affected U.S. firms and U.S. (George W. Bush) administration are quite upset about this especially since they were awaiting the outcome of OECD (Organization for Economic Cooperation and Development) negotiations on this issue. With this step, the EU is seen as having pre-empted this issue by its unilateral action. Moreover, the collection of the VAT by customer (home) state is going to cause more problems and uncertainties for American sellers than for their EU counterparts who can collect all such taxes for their home state at one rate. The matter is further complicated by the question of determining and verifying the actual country of residence for each customer. The given e-mail address of the customer may not be the real residential one if an attempt is made to circumvent the law by pretending to be living in a member state with a lower VAT rate. Or the customer may in effect be fronting for someone else—even reselling to consumers in higher VAT rate countries. The responsibility to tax the real customers at the correct rate is presumably the (non-EU) seller's. This could prove to be very difficult and quite costly for them. Some critics think that this in effect discriminates against non-EU firms and may contravene World Trade Organization rules on the liberalization of services trade. Another problem is that the different member states do not treat all activities the same way. For instance, AOL is subject to VAT for fees paid for Internet service in France and Germany but not in Britain (Maguire, 1999; Salak, 2000; Schwarz, 2000).

These rules only apply to business-to-consumer sales and not to business-to-business transactions. The latter are responsible for the bulk of the Web-based sales. Still, American online retailers selling to the EU will have some tough decisions to make on how to stay competitive, whether to absorb the tax or pass it along to the consumer and how to become more cost effective.

Income Taxes

Just like virtually any business local or foreign, e-commerce is required to pay (income) taxes on its earnings. There are three considerations here, viz. residence, source and nexus. American firms (i.e., those that have their residence or head office in the U.S.) are subject to U.S. income taxes on their worldwide income. Such businesses could be taxed on income in other countries if they earn money there. This could lead to so-called double taxation, meaning that the firm gets taxed twice (or in theory multiple times) by different countries for earnings originating from there. Fortunately, the U.S. has double taxation treaties with most of the industrialized countries, including the member states of the EU. This prevents businesses being taxed twice for the same income. These treaties are based on a model, which was designed by the OECD (Levin & Pedersen, 2000; Schwarz, 2000; Thibodeau, 1999).

For e-business, a particular concern is how their Web-based earnings are treated by foreign (in this case, EU) tax authorities. The OECD has determined that e-commerce is a digital delivery and, as such, is not any different from physical delivery. Whether a CD is physically shipped to or downloaded by the consumer makes no difference. In both cases, the earnings constitute business income which is NOT subject to tax in the source country. This means that for U.S.-based firms, no other country besides the U.S. is supposed to withhold (income) tax on this income.

Some e-businesses have a permanent establishment or nexus in another country (e.g., an office or warehouse and employees)–in short, a physical presence. What this means is that this portion of the business is most likely to be treated as a local firm for taxation purposes. (Please note that interpretations may differ–some hold that in many countries a simple warehouse and delivery operation does NOT constitute a permanent establishment.) Should such firms choose to use the facilities of other, say local EU firms, for this purpose, they will not be regarded as having a European nexus. So, this outsourcing is one way to legally avoid tax responsibility in another country.

The issue of nexus can be quite complicated, however. For instance, if a U.S. e-business has a Web site which is accessible in the EU, is that regarded as a permanent establishment? According to the OECD the answer is no since a mere virtual presence is insufficient. But what about using a specially-designed-for-the-EU Web site operated by an EU ISP? Or an interactive Web site specifically geared for the EU and staffed by EU personnel employed by an EU subcontractor? When does a virtual presence become a real or permanent one? This is an example of one of the many uncertainties and ambiguities in this field that still have to be resolved.

There is some merit in arguing that the entire concept of permanent establishment has become outdated with the advent of the Internet and the resulting easier access to global markets. Some experts are proposing that this concept be replaced by something more in tune with current developments. The problem is that they cannot seem to agree on what this should be. Even if they did, they would still have to sell it to the bureaucrats and politicians of numerous countries with different priorities–no mean task. This is one more example of the fluid state of affairs.

The dynamics of the Internet are also taking its toll on the concepts of *residence* and *source* since they are also both place related. Headquarters of e-businesses are often more mobile than their conventional counterparts and can consequently be moved more rapidly, such as offshore. Similarly, products or services delivered or downloaded in high-tax areas can often be easily directed elsewhere (e.g., with the aid of dummy or virtual customers).

Liability and Jurisdiction

When a U.S. online firm sells a product to an EU customer and the latter is injured by its use, what laws apply? Where does the consumer file suit? How can the firm defend itself? What happens if the suit is about an advertisement that is deemed offensive in a country (say Muslim) for which it was not intended? Can an American ISP be held criminally liable in France for Nazi paraphernalia sold on its U.S. Web site to a French customer who bypassed its French Web site because it did not contain these items in conformance with French law? How far can the (foreign) law reach? What if the laws or courts of different countries conflict or contradict each other?

These questions are a sampling of actual issues surrounding e-commerce. They create a very uncertain legal climate, one in which more and more firms are beginning to feel increasingly uncomfortable. No wonder some are pulling out of countries that they perceive as having a hostile or uncertain legal environment. The problem is that countries have failed to establish an international legal framework to protect the rights of buyers and sellers on the Internet. Attempts to create international treaties on such issues as liability, libel, and copyright infringements have failed. This means that the laws of the countries concerned will have to be used instead of global (treaty) laws. This, in turn, raises the question of which country's law applies in a given situation, bringing us to the issue of jurisdiction.

To start with, under U.S. contract law, one party may sue another in its home state (e.g., a consumer from Virginia may sue a seller in that state even if the seller is located in, say, California). As regards e-business, a tentative standard seems to be evolving in the U.S. where key standards are the degree of interactivity of the Web site and whether a sale took place. In cases where the site is purely informational, for instance, the company would most likely have to be sued in its own home state, but if a sale was made, the consumer's home state would most likely be the venue.

In contrast, the EU has initially emphasized the so-called home-country control principle. Under this doctrine, the seller's home country has jurisdiction. This view favors business. A few years ago, European consumer groups were successful in lobbying to reverse the EU directive so that currently European consumers can sue in their own country. This appears to bring it into line with general U.S. practice (Tucker & Eaglesham, 1999).

Other types of civil lawsuits include those for libel and copyright infringements. Based on limited case law so far, it would appear that defendants may be asked to appear in the courts of countries that were not even specifically targeted by the Internet activities of the defendant. Criminal prosecutions in foreign countries for violations, such as creating software or selling products that are perfectly legal in the home country but not in the prosecuting one, are also occurring (Newman, 2003).

The Hague Conference on Private International Law is attempting to draft an e-commerce treaty that would make court rulings in one country enforceable in all of the signatory countries. Skeptics say that the push for globalization of markets seen during the 1990s has lost its momentum. For various reasons including political, economic, and military (e.g., the Iraq war), countries tend to act more on their own and to assert their sovereignty. Many e-business and media firms that at first were very supportive of these treaties have backed off when they realized that such treaties would also incorporate laws from countries that would conflict with U.S. law (e.g., in the areas of free speech, liability, and copyright infringements). These areas failed to win consensus at the Hague Conference. This conference is now trying to reach agreement on a less controversial area, namely, how judgments will be enforced in business-to-business commercial contracts. These difficulties underscore the problems encountered when trying to bridge differences in the ethical, cultural, and economic values of the negotiating countries at a time when national issues often appear to enjoy a higher priority than globalization (Newman, 2003).

Marketing Implications: Opportunities and Challenges

Despite the difficulties mentioned above regarding regulatory issues, e-marketers nevertheless are presented with both opportunities and challenges. Following are two scenarios that present ways for online marketers to reach foreign markets. The first is using a Web site primarily for product information and image building, and the second is offering a full service Web site. Associated problems that may occur are highlighted based on the previous discussion.

A European consumer may visit an e-marketer's Web site in any given EU host country where it has such sites or the firm's home-country Web site may be contacted. The customer has several options: use it only to gather information through a product search, perhaps enter a chatroom/forum, or simply view advertisements. At this point, there are limited regulatory implications for the U.S. retailer in the EU. For example, if there is misleading advertising, it may be subject to EU and/or local country jurisdiction. Consumer feedback would be limited to the telephone or e-mail responses. Such communication may provide some useful consumer data to correct promotional problems or improve the Web site for the U.S. e-marketer. Since no actual transactions can be effected on such sites, the legal implications in the EU are limited.

Even with limited Web site presence, European governments are highly restrictive about marketing promotions. For example, Germany discourages comparative advertisements and "refused to allow a Land's End customer satisfaction guarantee to be used online" in order to protect smaller retailers in the home country (Siegel, 2004, p. 150). Another option for the European consumer is to visit the U.S. e-marketer's European full service local Web site. A full-spectrum marketing strategy allows the consumer to gather information, purchase a product or service, and receive accompanying customer service. However, such operations will most likely subject the e-marketer to local European regulations. This would include requiring prior permission from the customer to obtain, use, and disseminate any personal data and to inform the individual of such use.

Enhanced promotional restrictions now apply to sending online marketing material (i.e., opt-in requirement) as established by the recent European Privacy and Electronic Communications Directive in EU countries (i.e., replaces 1998 Directive). If e-mail contains personable and identifiable information, such as the recipient's name, the individual has the right to request the firm to stop (e.g., opt out right) sending such material. For business-to-business e-mail, the firms must not conceal their identity. Failure to comply with these new rules may result in some type of enforcement action. Note that the U.S. e-marketer's actual legal exposure resulting from these actions is vague at best, but the spirit of the new rule is clear (Robinson, 2003).

Furthermore, if the e-marketer has a local Permanent Establishment, for example, ships from a local (European) warehouse, it would likely be subject to VAT and income taxes due to local nexus. If, however, an independent server is used for this purpose, the exposure of the U.S. e-business may be limited due to its virtual presence.

Since the U.S. is currently less stringent regarding promotions and privacy issues, e-marketers may be more comfortable selling to customers in the EU through its home-country Web site. Although this will make it less likely that EU rules will apply, this possibility cannot be ruled out entirely. Due to the unresolved international issues between the U.S. and the EU, some EU rules might still be applied (e.g., through the Safe Harbor agreement on data privacy). The e-marketer may also be sued by a European customer in an EU member state court for a contractual dispute if jurisdiction in such a court was specified in the contract. Ultimately, most e-marketers will eventually have little choice but to rely mostly on Web sites within the EU adapted for local language, preferences, and so forth if they want to remain competitive.

Addressing Regulatory Issues and Future Trends

The regulatory challenges facing U.S. and e-marketers marketing in the EU are complex and often unresolved. To the dismay of U.S. and EU policymakers, the Internet also creates a new marketplace for illegal activity, such as the sales of unapproved new drugs or prescription drugs dispensed without a valid prescription through spurious online drug pharmacies. Global consumers may have difficulty identifying which sites sell legitimate products. For example, this is currently the case for cigarette sales online. The private sector has an important role in providing consumer education and in providing helpful information about the quality of products online.

To successfully conduct business under these circumstances requires careful planning aided by proper advice. Both business and legal issues have to be considered. As a starting point, the following short list of precautions is presented to enhance e-marketing efforts.

First, select Web sites (server location), warehouse locations, and any other possible manifestations of nexus carefully. Weigh the pros of improved customer service against the cons of greater liability and taxation. Also, select countries with zero or low VAT rates that favor your firm's product or service (e.g., digitized software, music, data, etc.), for instance, the U.K. for hardcover books (no VAT) and Luxembourg for various products (lowest rates for many products).

Second, carefully review and monitor the content and promotions on your Web site. Encourage free flow of information via advertisements to the Web visitor but ensure that advertisers agree to abide by rules of conduct. Provide only essential information as Web site content. Display any sponsorship and also give details of where the provider is based. Try to make the content on different Web sites as consistent and as compliant with the most common regulations as possible. Perhaps one of the more perplexing issues in the regulatory environment in the EU will be to seek ways to develop relationships with customers that engender trust. Studies indicate that consumers are apt to minimize the number of Web sites they purchase at and are influenced by Web sites that promote brand while they are seeking stable long-term relationships (Fassott, 2003). The challenge will be to strike a balance between communicating with the customer and collecting information in an unobtrusive way.

Third, as the number of laws, regulations, or enforcement actions increase, ensure that marketing practices are not vulnerable to local regulations. This applies not only to the Web site itself but to all other marketing activities (e.g., price discounts, promotional deals, and non-Web advertising). For example, Weblining is a hot-button issue in the U.S. between financial service firms and consumer activist groups. At the onset of the Internet, insurers and others firms using the Internet were accused of redlining or Weblining. Consumers who do not fit into a certain profile may not be offered a good or service and may be charged more (Danna & Gandy, 2002). Weblining now encompasses privacy issues especially in the ways that firms collect and analyze consumer data. It is clear from the U.S. experience that consumers should be informed of the ways information is collected about them.

Another example of the regulatory uncertainty surrounding many recently introduced online marketing practices relates to gambling or gaming across national borders. The crux of the matter is whether countries that allow some form of distance (including online) betting can prohibit consumers to participate in offshore gambling. This is presently the position of the U.S. and also of most member states of the EU. It is therefore ironic that recently major decisions by two key institutions have essentially reversed this position. The WTO found in favor of Antigua in its complaint against the U.S. because it prohibited U.S. residents operating from their home turf to participate in online gambling conducted from Antigua. It was found that such regulation unfairly impeded global trade in violation of WTO agreements. The U.S. is likely to appeal this decision which may delay the final outcome and its implementation for years.

The European Court of Justice held a similar opinion when a few months ago it ruled in two separate cases (Gambelli & Lindman, respectively) that both involved online gambling between residents of different member states. The court found that unjustified discriminating restrictions of online gambling between member states is an unlawful restraint of free (global) trade. Since the EU intends to take up this issue for possible ultimate legislation, while the U.S. appeal of its WTO case could drag on for some time, the final resolution of this issue will have to wait (Miller & Brinkley, 2004).

Finally, the antitrust problems of Microsoft serve as an illustration of another facet of online marketing regulation in the U.S. and the EU, respectively. The main thrust of its case deals with the use of its Windows software in computer hardware, which does not represent an online marketing practice, per se. However, the required incorporation of its Media Player in this software package to the exclusion of competing media software does represent a forced sale of services that could be marketed independently online by competitors. In this context, the antitrust rulings, in both the U.S. and the EU, have a direct bearing on online marketing. The EU's ruling is more severe, requiring virtually immediate action by Microsoft to offer a Windows version stripped of Media Player. Microsoft is expected to appeal this decision which means that this (antitrust) issue too, will not be clarified for a while (Kanter, 2004).

These examples all demonstrate the dynamic and still uncertain global regulatory environment in which e-commerce firms have to operate. Management needs to stay on top of these developments in order to formulate the most effective and legally acceptable online marketing strategies—a daunting task. It will require continuous feedback of pertinent information not only of the marketplace but also of related technological, political, and regulatory trends both at home and abroad.

Conclusions

The Internet provides new opportunities for e-marketers. In the past, exporting required significant investment in outlets and logistics. To engage in business with European countries, it was necessary to adhere to numerous local laws and regulations. Although e-commerce has eased some of these constraints, a viable entry strategy into the EU still demands cautious deliberation given newly emerging laws and practices. Such forethought may perhaps reduce the risk in unwanted legal entanglements. Unfortunately, applicable regulations are often difficult to identify due to uncertainty and controversy, both in the U.S. and in the EU. In fact, at present some areas do not appear to be covered at all.

The biggest challenge for U.S. e-marketers in all this is to determine which laws apply to a specific operation and to what extent it should pursue this quest. For instance, how far should it go to determine nexus given that in some cases different authorities in different and sometimes even within the same country may give different opinions on the matter? The next big challenge is how to formulate and implement marketing strategies that deal with these challenges in the most effective way.

References

Avlonitis, G.K., & Karayanni, D.A. (2000). The impact of Internet use on business-to-business marketing: Examples from American and European companies. *Industrial Marketing Management, 29*(5), 441-459.

Barlow, D. (1999, September). VAT aspects of e-commerce. *Business and Management Practice,* 40-43.

Danna, A., & Gandy, O.H., Jr. (2002). All that glitters is not gold: Digging beneath the surface of data mining. *Journal of Business Ethics, 40*, 373-386.

Fassott, G. (2003, June 23-25). *E-CRM tools and their impact on relationship quality and loyalty in e-tailing.* Proceedings of the International Customer Relationship Marketing Management Conference, Berlin, Germany.

Forrester Research, Inc. (2003). Retrieved October 25, 2004, from *http// www.forrestser.com*.

Grant, E. (2002, February 15). Battle brewing over European e-tail tax plan. *E-Commerce Times*. Hargreaves, D. (2000, May 10). Europe tries to clear ground for e-commerce: A directive seeks to clarify which national law applies to companies attempting to trade online. *Financial Times, 2.*

Kanter, J. (2004, September 26). Microsoft plots EU counterstrike. *The Wall Street Journal,* B4.

Levin, L.D., & Pedersen, R.C. (2000, May 31). When does e-commerce create a PE? *The Tax Adviser,* 306.

Maguire, N. (1999, September). Taxation of e-commerce. *Business and Management Practices,* 3-12.

Miller, S., & Brinkley, C. (2004, March 25). U.S. ban on Web gambling breaks global trade pacts, says WTO. *The Wall Street Journal,* A2.

Mitchener, B. (2003, November 10). European download blues. *The Wall Street Journal,* B1.

Newman, M. (2003, October 10). VAT's up? *The Wall Street Journal Online,* B1.

Robinson, D. (2003, November 24). E-marketing rules clarified. *Management Week,* p. 35.

Salak, J. (2000, July 10). A taxing situation – Who pays for what? Not even the recipients seem to know. *Telecom,* 52.

Schwarz, J. (2000, Jume 1). Virtual tax proves complex. *Financial Times,* 6.

Siegel, C. (2004). *Internet marketing: Foundations and applications.* Boston, MA: Houghton Mifflin.

Thibodeau, P. (1999, August 30). Europe wants U.S. firms to pay e-commerce taxes. *Computerworld, 33,* 35-41.

Tucker, E., & Eaglesham, J. (1999, July 6). Legislative threat to e-commerce. *Financial Times,* 2.

Utton, M.A. (2000, February). Books are not different after all. *International Journal of the Economics of Business,* 7(1), 115.

Waldmeir, P. (1999, December 23). Global e-commerce law comes under the spotlight. *Financial Times,* 9.

Wilson, S.G., & Abel, I. (2002). So you want to get involved in e-commerce. *Industrial Marketing Management, 31*(2), 85-49.

Zugelder, M.T., Flaherty, T.B., & Johnson, J.P. (2000). Legal issues associated with international internet marketing. *International Marketing Review, 17,* 2-3.

Section V:

E-Consumer
Theoretical Frameworks

Chapter XIV

Modeling the Effects of Attitudes Toward Advertising on the Internet

Chris Manolis, Xavier University, USA

Nicole Averill, Carmichael Events, A Division of Eight Entertainment, LCC, USA

Charles M. Brooks, Quinnipiac University, USA

Abstract

The notion that consumer attitudes toward products and advertisements affect consumer behavior is well-known in the marketing literature. Research and theory supporting this idea are based largely on traditional means of advertising (e.g., print ads, television commercials, etc.). Although research has begun to address if, when, and how the above process translates to Internet advertising, continued study is needed in this area. The following chapter depicts a study undertaken to further clarify the relationship between attitudes toward Internet advertising, brand attitudes, and purchase intentions. Given that various business models suggest future change and growth in sales and advertising revenues generated via the Internet, such a study is both necessary and valuable.

Introduction

Marketers have long been interested in the effects of advertising. Of particular interest is the relationship between a consumer's attitude toward an advertisement and how this attitude affects subsequent behavior (Shimp, 1981). Researchers have come to understand that, under certain conditions, consumer attitudes toward ads can ultimately affect purchase intentions and behavior (Brown & Stayman, 1992; Mitchell & Olson, 1981). Much of the knowledge and theory development in this area derives from research involving traditional means of advertising, such as print ads and television commercials (e.g., Lord, Lee & Sauer, 1995; MacKenzie, Lutz & Belch, 1986). Although a fair amount of research on attitudes toward Internet advertising has been conducted (Bruner & Kumar, 2000; Schlosser, Shavitt & Kanfer, 1999; Stevenson, Bruner & Kumar, 2000; Sukpanich & Chen, 1999), few if any studies have provided a straightforward experimental test of whether or not manipulated consumer attitudes toward Internet advertising affect brand attitudes and subsequent purchase intentions. This chapter represents such a study.

Internet advertising continues to grow in importance and has become a significant element of the marketing communications mix (Becker-Olsen, 2003). Some predicted that by 2007, Internet advertising will be the fourth largest advertising medium and that online advertising expenditures will reach $14 billion (Peach, Blank & Grahn, 2002). Although Internet advertising is growing, marketers are often unclear as to the effectiveness of this relatively new form of advertising (Dietz, 1998; Kennerdale, 2001) and question how the Internet compares to other more traditional marketing media (Jones, Pentecost & Requena, 2003).

The research set forth in this chapter addresses whether or not extant theory and research on the effects of attitude towards advertising are applicable to advertising on the Internet and purports to provide a straightforward empirical investigation toward this end. Based on the growing popularity of the Internet and the importance of knowing whether or not traditional advertising strategy and theory are applicable to Internet advertising (Maddox, Mehta & Daubek, 1997), a study is conducted. The study measures whether or not experimentally manipulated consumer attitudes toward Internet advertising affect brand attitudes and intentions to purchase. Before discussing Internet advertising further and ultimately offering research hypotheses, we first review the theory and research on consumer attitudes toward advertising.

Attitudes Toward Advertisements

Based primarily on the work of Fishbein (1967) and his associates (e.g., Fishbein & Ajzen, 1975), a consumer's attitude is thought to involve cognitive, affective, and conative components that combine such that a consumer's beliefs (cognitions) and feelings (affect) toward a given object (product or brand) or behavior create an attitude. This attitude, in turn, can ultimately affect the consumer's behavioral or purchase intentions (conative) toward the object (Lutz, MacKenzie & Belch, 1983; Mitchell & Olson, 1981). Since attitudes can determine purchase intentions, it is important that products and brands (as objects) elicit favorable attitudes.

In addition to forming attitudes based on products and product attributes (i.e., features), consumers often form distinct attitudes toward advertisements (A_{AD}) (Lutz, 1985; Mitchell & Olson, 1981). A_{AD} is thought to represent a predisposition to respond in a favorable or unfavorable manner to a particular advertising stimulus during a particular exposure occasion (Lutz, 1985). Given that consumer attitudes toward ads are potentially important, marketing researchers have investigated what causes advertising to elicit positive or negative consumer feelings (Lutz, 1985). Aaker and Stayman (1990), for instance, identified four factors that determine feelings toward advertisements. Positive feelings toward ads are elicited when the advertising is either informative or entertaining. Negative feelings, on the other hand, are elicited when advertising is either deceptive or irritating. Although not surprising, these findings provide advertisers with important insights toward developing successful advertising.

Attitudes toward ads are thought to play a significant role in the formation of brand attitudes, which, in turn, can affect purchase intentions (Biel, 1990; MacKenzie, Lutz & Belch, 1986). If consumers like a particular advertisement, they may be more likely to purchase the advertised brand by way of positive brand attitudes. Specifically, A_{AD} affects attitude toward the brand (A_B), which, in turn, affects purchase intentions (PI) (Mitchell & Olson, 1981).

Past Purchase Behavior and Subjective Norms

Attitudes are not the only factors that affect purchase intentions. A consumer's past purchase behavior (PPB) is also likely to have an impact on purchase intentions (Bentler & Speckart, 1979). For instance, consumers who have purchased particular products or shopped in particular stores or on particular Web sites in the past and had satisfactory experiences are likely to purchase those products or shop at those stores/sites again in the future (Bruner & Kumar, 2000). As will be seen shortly, we will focus on past Internet purchase behavior in the current study.

Another factor affecting purchase intentions is subjective norms (SN) (Ajzen & Fishbein, 1980). SN represent what someone perceives other people (significant other people) think of a particular behavior or, in a consumer context, what someone perceives other people think of purchasing a particular product or brand (Lutz, 1977). If reference group members (i.e., friends, relatives, etc.) are thought to approve of a behavior or brand (i.e., purchasing a particular brand), then the behavior or brand is more likely to be acceptable, and a consumer is more likely to purchase the brand. If, on the other hand, a brand is perceived as unacceptable by members of a reference group, a consumer is less likely to buy the brand (Figure 1).

Figure 1. Conceptual model of purchase intentions

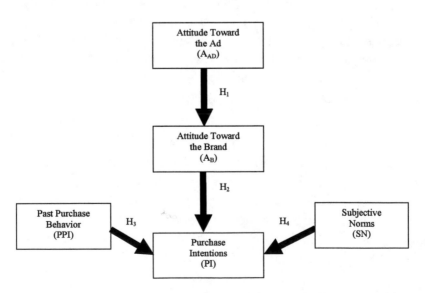

Internet Advertising

Broadly speaking, Internet advertising encompasses any form of commercial content available on the Internet that is designed by businesses to inform consumers about a product or service (Schlosser, Shavitt & Kanfer, 1999). This may include banner advertisements, e-mail advertisements, and corporate/organizational Web sites and/or home pages, all of which supply some depth of information about a company and/or its products (Singh & Dalal, 1999). This is a particularly important point currently as Internet advertising is conceptualized broadly in this chapter such that Internet advertising and Internet sites will be referred to and used almost synonymously (Ducoffe, 1996). Of course, appropriate and sufficient detail will be provided when necessary and/or appropriate in this regard.

The Internet represents one of the newest and potentially most powerful advertising mediums in the world. The Internet has grown in popularity as an advertising medium because, among other things, it allows 24-hour interactivity between the advertiser and customer. The Internet is also cost effective. With a Web site, unlike many traditional forms of advertising, the cost of reaching prospective customers does not increase proportionally with additional exposures (site visits). Internet advertising saves consumers time and allows them to obtain information regarding a wide variety of merchandise from multiple and convenient locations. Finally, Internet advertising has been said to facilitate the introduction of new products and services (Wolin, Korgaonkar & Lund, 2002).

From a marketing standpoint, firms need to establish a presence on the Internet in order to be competitive. Although many organizations are attempting to capitalize on this new medium, many do not market or advertise successfully their company or products on the Internet. Establishing an Internet presence does not necessarily imply success on the Internet. Like any advertisement, a Web site must possess positive qualities as evaluated by those coming in contact with the site, leading ultimately to positive attitudes and desired behavior. Advertising effectiveness on the Internet, therefore, would appear to be the result of sound marketing and creative strategy – much the same as successful advertising using traditional mediums.

Problems with the Internet as a Commercial Advertising Medium

Despite its proven popularity, the Internet and Internet advertising are not without their problems. Unlike traditional advertising, for example, Internet advertising is plagued by a relative lack of research addressing and measuring its effectiveness. For instance, it is still not entirely clear if successful strategies utilized in traditional advertising work for Internet-based advertising (Jones, Pentecost & Requena, 2003; Yoon, 2001). And, when research has been conducted in this area, Internet advertising – not unlike most advertising mediums – has proven difficult to measure as far as effectiveness is concerned (Becker-Olsen, 2003).

Some suggest that the Internet can pose distinct disadvantages as an advertising medium given that numerous studies have shown that certain factors result in consumers being irritated by the Web sites they visit. In one study, for instance, only three out of 100 sites studied were adequately servicing their visitors, and the other 97 either excluded or annoyed valuable and potential clients (*The Economist*, 1998). A study conducted by Dellaert and Kahn (1999) found that consumers experience negative or unpleasant feelings when they are confronted with long wait times. In a study concerning the intrusiveness of ads, Edwards, Li, and Lee (2002) found that consumers perceived pop-up ads to be irritating. Research done by INTECO, a Connecticut-based research firm, found that speedy downloads, along with reliable, up-to-date, and easy-to-find information are by far the most important Web site features according to consumers and that animation, sound, and video enhancement rate among the least important. Another study conducted by the Georgia Institute of Technology's *WWW Surveying Team* found that many Internet users perceive pages to be poorly designed and too slow to download.

Ratchford, Lee, and Talukdar (2003) note that evaluations of the Internet as a source of consumer information have been limited and argue that before making media and advertising-related choices, companies need to clearly understand the role of the Internet in consumer perceptions and behavior. Furthermore, the notion of interactivity (the idea whereby there are iterations between the firm and the customer, information from both parties is elicited, and attempts to align interests and possibilities are made) suggests that Internet advertising audiences – as compared to audiences of traditional advertising mediums – have the ability to choose and respond to ads of their liking thereby creating added obstacles and/or confusion on the part of Internet advertisers (Drèze & Hussherr, 2003).

Given that a large number of Web sites are doing things to potentially disappoint and/or displease consumers, it is important to understand precisely what consumers like and dislike with respect to commercial Web sites and Web-based advertising. As evidenced in these and other studies, three negative factors seem to be predominant with respect to the Internet as a commercial and advertising medium: (1) pages take too long to download, (2) sites are difficult to navigate, and (3) sites have too much irrelevant information. And, given the high costs associated with developing and maintaining an Internet site, companies must know whether extant strategies and measures associated with traditional advertising are applicable to Internet advertising.

Although there remains much to learn with respect to the Internet and Internet advertising, recent research in this area has amassed some insights regarding advertising on this new medium, particularly in the area of consumer attitudes. We turn now to some of these insights and research findings in order to position the current study and highlight its potential contribution to the literature.

Attitudes and Advertising on the Internet

Given that the current research spotlights how attitudes toward Internet advertising (Web sites, etc.) might affect brand attitudes (and subsequently how these attitudes might affect purchase intentions), it is important to review extant research that involves consumer attitudes and the Internet. In the following section, we review a sample of the more relevant work in this area.

First, it is important to understand that the study of consumer attitudes toward advertising on the Internet has emerged recently as a relatively significant area of Internet-related research and inquiry. In an effort to better understand what consumer attitudes toward advertising on the Internet are comprised of, as well as develop a scale useful in detecting such feelings, Sukpanich and Chen (1999) suggest that these attitudes likely involve awareness, preference, and intentions. In a related study, Wolin, Korgaonkar, and Lund (2002) discovered that Internet users' attitudes towards Internet advertising are a function of various belief factors. Product information, hedonic pleasure, and social role and image were positively related to subjects' attitudes toward Internet advertising, while materialism and falsity related negatively to these attitudes. The results also indicated that the more positive Internet users' attitudes toward Internet advertising, the greater likelihood the users would respond favorably to Internet ads (e.g., pay close attention to ads, click on ads to gather more information, etc.). Other research has attempted to actually measure consumer attitudes toward Internet-based advertising and discover whether or not such promotions are useful. Schlosser, Shavitt and Kanfer (1999), for instance, indicate that attitudes toward Internet advertising are generally positive and that a majority of users are at least somewhat confident in the validity of information they obtain via Internet advertising. Research by Ratchford, Talukdar, and Lee (2001) indicates that consumers rely on the Internet as an information source when making product choices. In perhaps the most pertinent study relating to the current research, Bruner and Kumar (2000) found that Internet experience – an experimentally manipulated independent variable – had a positive affect on five measured dependent variables: attitudes toward Web sites, attitudes toward commercials (commercial video clips playing on Web sites), attention paid to commercials, brand attitudes, and purchase intentions. It is also worth noting that the five dependent variables were positively associated.

In the end, the Internet represents a popular and complex if not confusing medium as far as advertisers and marketers are concerned. Accordingly, more research is needed to better understand consumers and their reactions and interactions with this new and powerful advertising medium. Although consumer attitudes toward Internet advertising have been addressed in the literature above, we contend that there is the need for a clear-cut experimental study examining how and if consumer attitudes regarding Internet advertising directly affect brand attitudes, which, in turn, affect purchase intentions. We further contend that such a study should necessarily adhere to extant attitude theory (Bentler & Speckart, 1979; Fishbein & Ajzen, 1975; Lavidge & Steiner, 1961; Mitchell & Olson, 1981; Shimp, 1981) such that the results of the study are straightforward. Next, we outline the objectives and potential contributions of the current study.

Research Hypotheses

This chapter addresses attitudes toward advertising as they apply to the Internet. Based on existing theory and past research, it is predicted that a consumer's attitude toward Internet advertising/a Web site (A_{AD} or $A_{Web\,site}$) will influence that consumer's attitude toward the brand being advertised on the Web site (A_B). This brand attitude will, in turn, influence consumer purchase intentions (PI). Purchase intentions will also be directly and positively affected by both the consumer's past purchasing behavior (PPB) and what the consumer perceives others think about buying the advertised brand (SN). We test the following hypotheses (see Figure 1).

H_1: *A positive attitude toward an Internet site (Internet advertising) will create a positive attitude toward a brand being advertised on the site.*

H_2: *A positive attitude toward the brand will increase purchase intentions for the brand.*

H_3: *The greater the extent of past purchase behavior, the more likely the individual is to purchase (intend to purchase) the advertised brand.*

H_4: *The more positive someone's perceptions about how others feel about purchasing the brand being advertised, the more likely the individual is to purchase the advertised brand.*

Method

Manipulation

To confirm that the consumer complaints discussed earlier are in fact some of the chief problems concerning Web sites (Web advertising), a pilot study was conducted. A convenience sample of 80 individuals who were familiar with the Internet were surveyed. The findings from the pilot study were consistent with previous research and indicated that the three most frequently cited Web site characteristics that irritated these Internet users were (1) pages with long download times, (2) irrelevant information on a Web site, and (3) an inability to navigate easily throughout a Web site. Additionally, respondents reported that the three features they appreciated most when visiting a Web site were (1) the Web site is informational, (2) the site is easy to use, and (3) graphics are used in addition to text. These findings also correspond to what Zhang and von Dran (2000) describe as *satisfiers* (factors and attributes in the Internet environment that contribute mainly to user satisfaction) and *dissatisfiers* (factors and attributes in the Internet environment that contribute mainly to user dissatisfaction). Based on the above mentioned attributes, both a good and a bad Web site were developed for the purposes of the study. Together, the sites will represent A_{AD} (attitudes toward advertising): the manipulated *independent variable* in the study.

Because sneakers represent a familiar product for most consumers, they were chosen as the focal product for these Web sites. The sneakers advertised on the two Web sites were a fictitious brand of running sneakers named *Zikes*. These shoes were said to be available over the Internet and were comparable in style and price to other brand name sneakers, such as Reebok and Nike. Unlike existing brands, however, *Zikes* were said to be manufactured to custom fit each individual's feet. When consumers went to the *Zikes* Web site, they found a footmap that they could print out. Having printed out the footmap, consumers could trace their feet onto the map. Consumers could select sneaker style and color, and, when *Zikes* (the company) received payment and the footmap, a custom pair of the sneakers would hypothetically be produced. The footmap was used to differentiate the *Zikes* brand from those of the competition and to get consumers interested and involved in the product and thereby the study. Since the objective of the study was to determine whether a Web site/ Web site advertising could affect a consumer's attitude toward a product/ brand, it was necessary to make the product advertised on the site as interesting as possible.

The good Web site. The good *Zikes* Web site contained five Web pages. Each page downloaded relatively quickly, contained only information relevant to *Zikes* sneakers, and was easy for subjects to navigate from one page to the next via a navigation menu at the bottom of each page. The home page featured a picture of a *Zikes* sneaker, along with an informative paragraph about the product. Four navigation buttons, About Us, Sneaker Styles, Footmap, and Order Form, were located under the copy.

Clicking on the About Us button took users to a page consisting of a graphic at the top of the page which read "About *Zikes* Running Shoes" followed by a narrative explaining the history of the company. When users clicked on the Sneaker Styles button, they were taken to a page consisting of a graphic at the top of the page which read "Sneaker Styles," followed by six two-inch by one-inch pictures of the various sneaker styles with a description of features, colors, and prices. As with other pages, there were four navigation buttons at the bottom of this page. When consumers clicked on the Footmap page, they were taken to a page consisting of a printable graphic at the top of the page which read "Footmap," as well as a picture of a grid with a faint outline of a life-sized foot. Once again, under the pages were the four navigation buttons, the last of which was the Order Form button. When the consumers clicked on the Order Form button, they were taken to a page consisting of a graphic at the top of the page which read "Order Form" and a traditional order form for customers to complete.

To summarize, three key features were built into the Web site that made it good from a promotional standpoint. First, the layout of each page was consistent and allowed the user to navigate easily. Second, each page provided an adequate amount of information relevant to *Zikes* sneakers but did not include extraneous information that did not pertain directly to the product. Specifically, there were no additional links to other sites, there were no advertising banners for other companies, and there were no broken links. Finally, the graphics and text used on this site were sized appropriately such that download times were relatively fast. Since features that most commonly irritate Internet-users/ consumers were avoided, and the features that consumers generally find pleasing were included, this site was designated as the good site/advertisement.

The bad Web site. The bad Web site was designed to be the opposite of the good Web site. Although the core informational content was similar across the two sites, various pages on the bad site contained additional promotional information that did not directly pertain to *Zikes*. This, in turn, resulted in longer download times. Although the home page on the bad site contained the same introductory paragraph as the home page on the good site, the bad site page also contained pictures of all six sneakers made by *Zikes*. Three of these pictures linked to sneaker styles page (where the sneakers were described) and three led to dead pages that resulted in error messages. The page also featured banner advertisements for running-related magazines (e.g., *Track and Field* and *Runners' World*) that also led to error messages. There was an advertisement for an online bookstore that featured books about running; when the users tried to access the bookstore page they again encountered an error message. Finally, there was a Featured Article column where users could access articles about running.

As with the good site, the bad site was designed such that navigation was through a series of menu buttons. Unlike the good site, however, the bad site only contained two (compared to four) navigation buttons per page. In order for consumers to access the other *Zikes* pages they had to bounce from page to page, and access to the links was through another page altogether. When consumers/subjects clicked on the Footmap button, they were led to the same footmap as on the good site, but the only navigation button on the footmap page was the Home Page button. The sneaker styles page was the same as on the good site, except that the promotional pictures were larger causing longer download times, and the descriptions were not as in-depth as on the good site. The only way to access the order form was via the home page. Although the overall product information was consistent across the sites (advertisements), the bad compared with the good site contained less relevant promotional information, contained more irrelevant information, was more difficult to navigate, and had slower download times.

Procedure

Subjects were recruited from various undergraduate marketing courses at a university in the northeastern United States. Student subjects were told that they were going to be part of a new product testing group and that the new product being tested was a new line of sneakers called *Zikes*. Subjects were given the following information:

"The purpose of this research is to investigate different methods of pre-testing a new product which is still in the "conception stage" of development. Your task is simply to examine the following website, and, subsequently, form an evaluation of the new product, Zikes Sneakers."

This introduction was adapted from a study done by MacKenzie and Lutz (1989). In short, the students were led to believe that they were part of a new product development study and were unaware that the researchers were interested in measuring their Web site-based, promotion-related attitudes.

After the introduction, subjects were randomly given a sheet of paper with instructions and a Web address where one of the *Zikes* Web sites was posted. Each subject visited *either* the good or the bad Web site. As mentioned previously, these sites or advertisements represented the A_{AD} (attitude toward advertising or Internet advertising) *manipulated independent variable* in the study. Subjects were instructed to spend about 10 minutes exploring the site. They were asked to access and evaluate all the pages, links, and information on the site. After exploring the site, subjects were asked to complete a questionnaire pertaining to their site visit.

One hundred sixty student subjects were included in this main study, and approximately equal numbers of both males (49%) and females participated. The limitations associated with the use of student samples are well-known and based largely on the belief that such samples do not necessarily represent the general population (e.g., Ferber, 1977). There have recently been numerous assertions made in the Internet/advertising literature, however, suggesting that undergraduate university students are particularly relevant for studies where the Internet is concerned as these students are typically young (as a relatively new medium, the Internet is still used primarily by younger people) and have a lot of experience with the medium (Danaher & Mullarkey, 2003).

Questionnaire

A questionnaire was designed to measure two additional *independent variables,* SN (subjective norms) and PPB (past purchase behavior), and the two *dependent variables* or constructs in the study, A_B (attitude towards brand) and PI (purchase intentions) (Figure 1). As each of these variables represent established and valid constructs, the majority of items on the survey either came directly from or were based on pre-existing reputable scales and studies. All items followed a Likert-type response format of 1 = Strongly Disagree and 7 = Strongly Agree.

The nine items measuring A_B represented ratings on how the subjects felt about the brand advertised on the Web site (e.g., "I think *Zikes* sneakers are a good product idea.") and are based on a study by Lutz, MacKenzie, and Belch (1983), among others.[1] The three items measuring SN (e.g., "People who are important to me would approve of me purchasing these sneakers.") were taken from studies by Bentler and Speckart (1979) and Lutz (1977). The PPB variable was also based on work by Bentler and Speckart (1979), yet modified for the current study. We asked subjects three questions pertaining to their past Internet purchase behavior given the study's Internet focus (e.g., "I have purchased products via the Internet on a regular basis over the past year."). In order to provide a more complete or comprehensive test of the model, the PI construct was conceptualized accordingly to three related yet distinct components: General Purchase Intention (GPI) (e.g., "This is a product that interests me enough that I would consider purchasing it."), Retail Purchase Intention (RPI) (e.g., "If given the opportunity, it is unlikely that I would purchase *Zikes* sneakers from a retail outlet." [reverse-coded]), and Internet Purchase Intention (IPI) (e.g., "I would be interested in purchasing *Zikes* sneakers off the Internet."). The items used to measure these PI components were based on work by Darley and Smith (1993), Bentler and Speckart (1979), and Lutz,

MacKenzie, and Belch (1983). Both the GPI and RPI variables consisted of two items each, while the IPI measure utilized a single item.

In order to verify the reliability of the constructs, Cronbach's coefficient alphas were computed for each construct comprised of more than two items. Coefficients ranged from 0.85 to 0.91. For the two constructs measured via two items each (GPI and RPI), a correlation coefficient (r) was computed. The r's for the items comprising GPI and RPI were 0.61 and 0.36 (p's < .05), respectively.

Results

Path Analyses

Figures 2 through 4 depict relationships between A_{AD} (the good versus the bad Web site) A_B, PPB, SN, and PI. According to the hypotheses, the models posit that a consumer's attitude toward the Web site/Web site advertising (manipulated independent variable) should affect that consumer's attitude toward the

Figure 2. Path analysis for general purchase intention

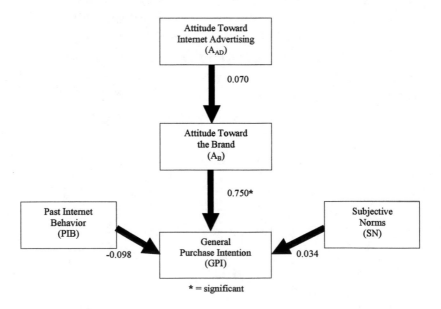

Figure 3. Path analysis for retail purchase intention

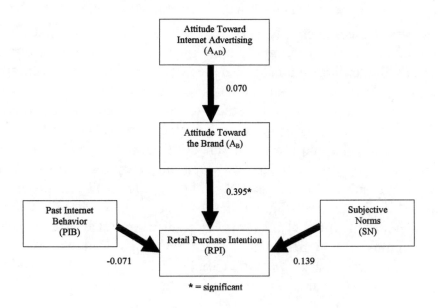

Figure 4. Path analysis for Internet purchase intention

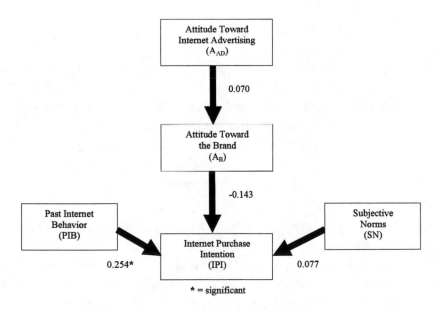

brand, which, in turn, should affect various intentions to purchase the product. Additionally, the consumer's past Internet purchase behavior as well as subjective norms should influence purchase intentions.

Path analysis was conducted using Bentler's EQS program. Parameter estimates from the analyses are reported in Figures 2 through 4. Contrary to predictions, the findings indicated that the Internet or Web site advertising attitudes (A_{AD}) had no influence on attitudes toward the brand (A_B). Hypothesis one is not supported. As expected, attitudes toward the brand did significantly affect purchase intentions, general and retail intentions (GPI and RPI, respectively) specifically. Attitudes toward the brand did not, however, affect Internet purchase intentions (IPI). Hypothesis two is partially supported. Also contrary to predictions, subjective norms (SN) and past purchase behavior on the Internet (PPB) did not significantly affect general or retail purchase intentions. Past behavior on the Internet did, however, significantly affect Internet purchase intentions. Although hypothesis four is not supported, hypothesis three is partially supported. Lastly, we tested the indirect effect of A_{AD} on the various purchase intention measures and found no significant effects.

Discussion

Attitudes Toward the Ad

A notable if not surprising finding in the current study reveals that attitudes toward Internet advertising did not impact brand attitudes. This finding is generally inconsistent with theory and previous research pertaining to traditional forms of advertising where attitudes toward advertisements have been shown to impact attitudes toward brands. The finding suggests that perhaps the theories or models pertaining to attitudes toward advertising (Lutz, MacKenzie & Belch, 1983; Mitchell & Olson, 1981) are not necessarily (or currently) applicable to the Internet.

The Internet (or Web sites) compared with traditional advertising represents a fundamentally different medium. Because the Internet and Web sites represent a new advertising medium, poorly designed Web sites/advertisements have arguably been the norm and well-crafted Web sites/advertisements the exception. As such, consumers may be accustomed to poorly designed Internet advertisements such that the impact on attitudes toward brands due to poorly designed Web sites/advertisements is minimal. In contrast, traditional advertising (e.g., print ads, TV commercials, etc.) is far more developed and a poor ad stands out thereby negatively impacting brand attitudes. In addition, consumers have more control over information presented on the Internet (Drèze & Hussherr, 2003). If this is indeed the case, it is reasonable to assume that this situation will not last indefinitely and that, at some point, consumer Internet advertising standards or expectations will approach those associated with traditional forms of advertising. For the time being, however, it is perhaps easier to avoid negative aspects of Internet advertising compared with more traditional advertising mediums.

A related explanation suggests that this null finding is the result of an inadequate experimental manipulation (i.e., good versus bad Web site). Despite our best efforts (including a pilot test and adherence to extant literature and research findings), it may be that the two contrasting Web sites/advertisements did not invoke the desired effect (i.e., different attitudes).

Purchase Intentions

As per attitude theory and its application to consumer behavior, purchase intentions represent a critical outcome variable or construct in the current study. Although the study and corresponding manipulation focused on the Internet, the brand in question (*Zikes* sneakers) could conceivably have been purchased (intended to be purchased) via numerous non-Internet channels. Accordingly (and as mentioned earlier), we utilized three components or versions of the purchase intentions construct (general, retail, and Internet purchase intentions) in order to provide a more inclusive test of the model. In this way, we hoped to provide insight into whether or not Internet purchase intentions are the only type of purchase intentions affected by past Internet purchase behavior. Alternatively, perhaps past Internet purchase behavior also affects general- and/or retail-related purchase intention measures. Another potential insight includes whether or not brand attitudes formed as a result of brand exposure via the Internet (only) affect varying purchase intention measures the same way.

General and retail purchase intentions were, as expected, affected positively and significantly by attitudes toward the brand. In support of extant theory, one's intention to purchase a particular brand—in general and at a retail outlet specifically—are at least partially based on the customer's attitude toward the brand. This finding highlights the importance of consumer brand attitudes as per a firm's bottom line. These purchase intention measures were not, however, affected by either subjective norms or past Internet purchase behavior. Although interesting in their own right, the insignificant results involving the past Internet purchase behavior variable are not entirely surprising given the specificity of this past purchase behavior measure. The insignificant effects pertaining to the subjective norms variable are, however, somewhat curious and are perhaps the result of the hypothetical or contrived nature of the (experimental) purchase setting.

Results from the study indicate that one's future online purchases (purchase intentions) are best predicted by one's past online purchasing behavior. Apparently, the effects of past Internet purchase behavior are limited to future Internet purchases. Furthermore, one's attitude toward a brand does not affect Internet purchase intentions. Given the fact that subjects' only affiliation with the *Zikes* brand was via the Internet (Web site), it is conceivable that the brand itself was subsumed under a commanding and comprehensive Internet purchase reaction or feeling. That is, perhaps the Internet purchase intention measure reflected the general willingness of subjects to purchase products via the Internet and, as such, was not affected by attitudes toward the specific brand. Finally, it is worth noting that this was the only single-item purchase intention measure which may help to explain the lack of significant findings regarding this outcome variable.

Implications

The results of this study have significant implications, both theoretically and managerially speaking. It is important to recall that the basic or core content of both Web sites (good versus bad) was identical. Despite every effort to create a good and bad Web site/advertisement, the sites did not prove to be significantly different. It appears, therefore, that perhaps the product or brand itself essentially dominated respondents' reactions to the Web site advertising at least as far as general and retail purchase intentions are concerned. Regardless of the precise mechanism by which the present findings occurred, the results do suggest that perhaps extant or traditional consumer advertising theory is not generally or necessarily applicable and/or accurate when it comes to the Internet. And, it is unclear whether this apparent disparity is temporary or more fundamental in nature. This is certainly an area for future research.

From a managerial standpoint, the findings suggest that actual products or brands are of the utmost importance when utilizing Web-based Internet advertising media. Weak Web site advertising showcasing a satisfactory product or brand may cause minimal damage. Conversely, strong or effective Web site advertising may not compensate for a poor product or brand. As a first priority, managers should focus on procuring a successful well-liked product or brand before focusing on what is said about a product or brand (and/or how the message is communicated) on the Internet. Although the Internet is said to facilitate the introduction of new products and services (Wolin, Korgaonkar & Lund, 2002), from an advertising perspective, this new medium may not be the best choice as far as facilitating initial positive product or brand attitudes are concerned.

The impact of Web site advertising will likely evolve over time. As effective Web site advertising becomes the norm versus the exception, the impact of poor Web site advertising on attitudes toward brands and purchase intentions is likely to increase. Future research needs to monitor evolving relationships between Web site advertising, attitudes toward brands, and purchase intentions.

Conclusions

In summary, there are two important findings reported in this chapter. First, it was found that Web site advertising does not appear to significantly impact attitudes toward a brand. This finding is contrary to previous research addressing attitudes toward advertisements. Further research is needed to verify this finding. Second, although attitudes toward the brand affected general- and retail-based purchase intentions, subjective norms and past Internet purchase behavior did not. In addition, Internet purchase intentions were only affected by a consumer's past purchasing behavior on the Internet. Although interesting, these findings are exploratory in nature and ongoing studies are needed.

References

Aaker, D.A., & Stayman, D.M. (1990, September). Measuring audience perceptions of commercials and relating them to ad impact. *Journal of Advertising Research*, 7-17.

Ajzen, I., & Fishbein, M. (1980). *Understanding attitudes and predicting social behavior*. Englewood Cliffs, NJ: Prentice Hall.

Becker-Olsen, K.L. (2003). And now a word from our sponsor. *Journal of Advertising, 32*(2), 17-33.

Bentler, P.M., & Speckart, G. (1979). Models of attitude–behavior relations. *Psychological Review, 86*(5), 452-464.

Biel, A.L., (1990, September). Love the ad, buy the product? *Admap*, 21-25.

Brown, S.P., & Stayman, D.M. (1992). Antecedents and consequences of attitude toward the ad: A meta-analysis. *Journal of Consumer Research, 19*(1), 34-51.

Bruner, G.C., & Kumar, A. (2000, January/April). Web commercials and advertising hierarchy-of-effects. *Journal of Advertising Research, 40*, 35-42.

Danaher, P.J., & Mullarkey, G.W. (2003, September). Factors affecting online advertising recall: A study of students. *Journal of Advertising Research, 43*, 252-267.

Darley, W.K., & Smith, R.E. (1993). Advertising claim objectivity: Antecedents and effects. *Journal of Marketing, 57*, 100-113.

Dellaert, B.G.C., & Kahn, E.E. (1999). How tolerable is delay? Consumers' evaluations of Internet Web sites after waiting. *Journal of Interactive Marketing, 13*(1), 41-54.

Dietz, N. (1998). Survey: Banners losing effectiveness. *Advertising Age's Business Marketing, 83*(9), 40.

Drèze, X., & Hussherr, F.X. (2003). Internet advertising: Is anybody watching? *Journal of Interactive Marketing, 17*(4), 8-23.

Ducoffe R.H. (1996, September-October). Advertising value and advertising on the Web. *Journal of Advertising Research, 36*, 21-35

The Economist (1999, February 28). Working the Web. *346*(8057), 67.

Edwards, S.M., Li, H., & Lee, J. (2002). Forced exposure and psychological reactance: Antecedents and consequences of the perceived intrusiveness of pop-up ads. *Journal of Advertising, 31*(3), 83-95.

Ferber, R. (1977). Research by convenience. *Journal of Consumer Research, 4*(1), 57-58.

Fishbein, M. (1967). Attitudes and the prediction of behavior. In M. Fishbein (Ed.), *Attitude theory and measurement* (pp. 477-492). New York: John Wiley & Sons.

Fishbein, M., & Ajzen, I. (1975). *Belief, attitude, intension and behavior: An introduction to theory and research*. Reading, MA: Addison-Wesley.

GVU's 9th WWW User Survey. (1997). Retrieved October 25, 2004, from *http://www.gvu.gatech.edu/user_surveys/survey-1998-04/*

Jones, M.Y., Pentecost, R., & Requena, G. (2003). Memory for advertising and information content: Comparing the printed page to the computer screen. *Advances in Consumer Research, 30*, 295-297.

Kennerdale, C. (2001). Banner advertising: Still alive, but fundamentally flawed? *EContent, 24*(2), 56-58.

Lavidge, R.J., & Steiner, G.A. (1961). A model for predictive measurements of advertising effectiveness. *Journal of Marketing, 25*, 59-62.

Lord, K.R., Lee, M., & Sauer, P.L. (1995). The combined influence hypothesis: Central and peripheral antecedents of attitude toward the ad. *Journal of Advertising, 24*(1), 73-85.

Lutz, R.J. (1977, March). An experimental investigation of causal relationships among cognition, affect, and behavioral intentions. *Journal of Consumer Research, 2*, 197-208.

Lutz, R.J. (1985). Affective and cognitive antecedents of attitude toward the ad: A conceptual framework. In L.F. Alwitt & A.A. Mitchell (eds.), *Psychological processes and advertising effects* (pp. 45-63). Hillsdale, NJ: Lawrence Erlbaum.

Lutz, R.J., MacKenzie, S.B., & Belch, G.E. (1983). Attitude toward the ad as a mediator of advertising effectiveness: Determinants and consequences. In R.P. Bagozzi & A.M. Tybout (eds.), *Advances in consumer research* (pp. 532-539). Ann Arbor, MI: Association for Consumer Research.

MacKenzie, S.B., & Lutz, R.J. (1989, April). An empirical examination of the structural antecedents of attitude toward the ad in an advertising pretesting context. *Journal of Marketing, 53,* 48-65.

MacKenzie, S.B., Lutz, R.J., & Belch, G.E. (1986, May). The role of attitude toward the ad as a mediator of advertising effectiveness: A test of competing explanations. *Journal of Marketing Research, 23,* 130-143.

Maddox, L.M., Mehta, D., & Daubek, H.G. (1997, March-April). The role and effect of Web advertising in advertising. *Journal of Advertising Research, 37,* 47-59.

Mitchell, A.A., & Olson, J.C. (1981, August). Are product attribute beliefs the only mediator of advertising effects on brand attitude? *Journal of Marketing Research, 18,* 318-332.

Peach, A., Blank, J., & Grahn, R. (2002). Online advertising and e-mail marketing through 2007. *Jupiter Research, Digital Content Vision Report, 2*(16).

Ratchford, B.T., Lee, M., & Talukdar, D. (2003). The impact of the Internet on information search for automobiles. *Journal of Marketing Research, 40*(2), 193-209.

Ratchford, B.T., Talukdar, D., & Lee, M. (2001). A model of consumer choice of the Internet as an information source. *International Journal of Electronic Commerce, 5*(3), 7-21.

Schlosser, A.E., Shavitt, S., & Kanfer, A. (1999). Survey of Internet users' attitudes toward Internet advertising. *Journal of Interactive Marketing, 13*(3), 34-54.

Shimp, T.A. (1981). Attitude toward the ad as a mediator of consumer brand choice. *Journal of Advertising, 10*(2), 9-15.

Singh, S.N., & Dalal, N.P. (1999). Web home pages as advertisements. *Communications of the ACM, 42*(8) 91-98.

Stevenson, J.C., Bruner, G.C., & Kumar, A. (2000). Webpage background and viewer attitudes. *Journal of Advertising Research, 40*(1/2), 29-34.

Sukpanich, N., & Chen, L.D. (1999, April). Measuring consumers' attitudes to Web advertising. *Electronic Markets, 9,* 20-24.

Wolin, L.D., Korgaonkar, P., & Daulatram, L. (2002). Beliefs, attitudes and behaviour towards Web advertising. *International Journal of Advertising, 21,* 87-113.

Yoon, S.J. (2001). What makes the Internet a choice of advertising medium? *Electronic Markets, 11*(3), 155-162.

Zhang, P., & von Dran, G.M. (2000). Satisfiers and dissatisfiers: A two-factor model for Website design and evaluation. *Journal of the American Society for Information Science, 51*(14), 1253-1268.

Endnotes

[1]   Although the original Fishbein theory (Fishbein & Ajzen, 1975) suggests that attitudes toward behaviors compared to attitudes toward objects and brands are likely better predictors of behavioral intentions, there is empirical evidence and literature in consumer behavior to suggest that attitudes toward brands are satisfactory in attitude models inclusive of the attitudes toward advertising construct (see Lutz, MacKenzie & Belch, 1983; MacKenzie & Lutz, 1989; Mitchell & Olson, 1981).

Chapter XV

Virtual Community:
A Model of Successful Marketing on the Internet

Carlos Flavián, University of Zaragoza, Spain

Miguel Guinalíu, University of Zaragoza, Spain

Abstract

The development of commerce on the Internet based on virtual communities has become one of the most successful business models in the world today. In this chapter, we analyze the concept of the virtual community, describe marketing strategy implications and guidelines for its management, and suggest some alternative strategies and technologies which could be used in running a virtual community. Thus, we attempt to discover the potential of virtual communities for an Internet entrepreneur, from a basically practical perspective, without losing sight of the conceptual aspects defining the strategies explained.

Introduction

When the technological bubble burst at the beginning of 2000, many companies in existence three years ago disappeared (Webmergers, 2003). This phenomenon has created the need to carefully analyze the economic viability of Internet projects. In this respect, researchers have concentrated fundamentally on defining aspects, such as consumer behavior or the design of more efficient business models and marketing strategies (Alba et al., 1997; Bauer, Grether & Leach, 2002; Bigné & Ruiz, 2003; Clemente & Escriba, 2003; Görsch, 2001; Ruiz & Bigné, 2003). A simple analysis of the present situation on the Web shows that many of the businesses which have survived the Internet crisis have a common link, namely, the existence of a virtual community organized around the entity (e.g., eBay, Wired, Yahoo!, Amazon). The virtual community, a social institution which goes back a very long way, has transferred to the Internet with considerable speed and ease and has served as a source of income for and is the mainstay of many organizations.

The general aim of this chapter is to explain the characteristic features of virtual communities, paying special attention to the most important strategies, management suggestions, and technology being used in this type of business model. This chapter begins with an analysis of the concept of community from a sociological viewpoint. It then deals with the definition of a virtual community and what causes an individual to belong to one. We shall then show the main implications of the development of virtual communities on marketing. Next, we detail the most feasible strategies and a series of recommendations for the proper management of Internet communities. There follows an explanation of the most common technologies in this type of business. The final section presents the main conclusions of the chapter and future lines of research and challenges to be addressed.

Analysis of the Virtual Community

In the field of sociology, a wide variety of communities have been defined on the basis of various criteria. Nevertheless, in this work, we propose to focus our attention on analyzing communities which have developed around a brand, in view of the fact that this is the type of community that has the greatest business potential and implications for marketing. Specifically, we shall consider that a brand community is a relationship structure built on admiration for a particular brand (Muñiz & O'Guinn, 2001). From a basically sociological perspective, Muñiz and O'Guinn (2001) consider that the brand community may be defined as having three principal elements:

1. *Consciousness of kind*: This aspect refers to the common interest shared by the community members, and, therefore, this is the most important factor to consider when defining a community. Consciousness of kind is the sentiment which sustains the cohesion of the group, as it is this feeling which links the individual to the other members and to the brand.

2. *Rituals and traditions*: These are processes which serve to reinforce consciousness of the brand and improve instruction in communal values. In general, these processes refer to the particular behavior of members of the community.

3. *Sense of moral responsibility*: This reflects the sentiments by which the members of a community feel morally obliged. This moral obligation implies that the community member should act by recruiting new members and retaining existing ones.

Focusing on the virtual community, it may be said that the first virtual community was actually created in the 1970s, specifically with the Talkomatic software, designed by Doug Brown of the University of Illinois in the autumn of 1973. However, at the time, its use was restricted to the scientific and military sector (ARPANET network).

It was not until the 1990s that there was an exponential growth of this type of organization. Available literature provides several definitions of the concept of virtual community. Among the most direct and functional, we would draw attention to the approach of authors, such as Cothrel and Williams (1999), who define it as a group of individuals which uses computer networks as a form of primary interaction, or Kardaras, Karakostas, and Papathanassiou (2003), who consider it to be a group of individuals who communicate by electronic means, such as the Internet, who share interests without needing to be in the same place or having physical contact or belonging to a particular ethnic group. From their members' perspective, virtual communities offer various benefits, prominent among which are the following (Hagel & Armstrong, 1997):

- *Dealing with matters of interest*: Virtual communities enable individuals to share information on matters of interest. For example, iVillage (*http://www.ivillage.co.uk*), a Web site owned by Tesco's Internet distribution division, offers a large number of services for women, such as advice on health, beauty, work, and leisure.

- *Establishing relationships*: In virtual communities, it is possible to encounter friendships, romantic relationships, or other users with similar experiences (e.g., http://www.friendster.com for the growth in number of friendships).

- *Living out fantasies*: In virtual communities (e.g., *http://www.cokemusic.com* or the online game GemStone, *http://www.play.net/gs4/new.asp*), members are able to share their different experiences (visit imaginary worlds, play, talk, etc.) in real-time and through virtual characters.

- *Carrying out transactions*: In virtual communities, individuals may also carry out commercial transactions, such as the case of Manchester United, which has realized that the Internet is a way of increasing sales of tickets and other licensed products of the club, as well as improving relationships with fans (*http://www.manutd.com* and *http://www.shop.manutd.com*).

Virtual Communities and Marketing Strategy Implications

Exploitation of virtual communities implies several marketing strategy implications are discussed.

New Forms of Communication (Viral Marketing)

The success of many of the initiatives based on the development of virtual communities is a true reflection of the changes caused by the Internet in marketing activities. These changes, explained by authors such as Hoffman and Novak (1996), typically consist of the replacement of traditional communication models (one-to-many) with others in which the interactions between all the participants in the market are constant (many-to-many). The virtual community represents these changes since it uses the Internet to make frequent contact among group members. This virtual network of contacts generates a new milieu in which marketing activities are partly reorientated, replacing massive advertising campaigns with others that make use of the potential of word-of-mouth transmission; this is known as viral marketing. As a result of viral campaigns, messages issued by the company to the community will, within a short period of time, reach all the members of the group, obviating the need for costly investments in the mass media and thus reducing promotional costs (Barnatt, 2001; Guthrie, 2000; Kardaras et al., 2003; Wang, Yu & Fesenmaier, 2002).

Information Source

The virtual community becomes a relevant information source for the making of decisions (Hagel & Singer, 1999). It enables customers to be involved in the development of new products, improves relationships, and promotes understanding of consumer behavior (Barnatt, 2001; Holmström, 2000; Kardaras et al., 2003).

Creation of Barriers Preventing the Entry of New Competitors

The virtual community creates barriers against new competitors and enables comprehensive individual marketing (Barnatt, 2001; Hagel & Armstrong, 1997; Kardaras et al., 2003).

Increased Security and Trust

Belonging to a virtual community calls for its members to comply with the norms that the group has imposed on itself. This self-regulation brings about a situation whereby those in the community meet their commitments, which in turn brings about an increase in trust. Thus, the community creates a climate of security based on reputation, which complements or even outdoes the most sophisticated technological systems (Klang, 2000). This phenomenon may be clearly seen in online auction communities (e.g., eBay) and in communities with risk-management systems based on reputation (Dellarocas, 2001; Kollock, 1999). There are two types of risk-management mechanisms, depending on whether they are based on the negative or positive reputation of the individual. With those based on negative reputation, the community members who feel that they have been cheated by the behavior of another member make their feelings known, usually via messages in the forum or the drawing up of blacklists. In the second instance, we have mechanisms based on positive reputation where satisfaction with the behavior of the other party is made public. In this case, there are various procedures, such as ratings for the individuals taking part. This system is based on the rating that one user gives another in terms of how the transaction has gone. When an individuals have high ratings, they have an incentive to maintain it, since it enables them to charge higher prices in auctions. In addition, this increases the security and trust of those participating in these auctions as they are trading with individuals with good standing.

Improved Capacity to Compete Against Traditional Companies

Traditional companies that have an online division (known as brick-and-mortar companies) have certain competitive advantages over organizations that distribute their products exclusively over the Internet (known as dot-com companies). These advantages, the result of the existence of inter-channel synergies, appear in aspects such as a greater degree of trust by individuals arising from a physical presence (Steinfield et al., 2002). The virtual community, as we shall now see, has the ability to overcome this type of advantage and is thus a good tool for increasing the company's competitive situation. According to certain authors (Guthrie, 2000; Wang et al., 2002), virtual communities are an ideal framework for implementing relationship marketing strategies.

Becoming a Source of Indirect Income

The business models of many organizations currently functioning on the Internet are characterized by the fact that they are almost exclusively financed by income from advertising (for example, digital newspapers, such as *The New York Times*, *http://www.nyt.com* or search engines, such as Google, *http://www.google.com*). In the case of virtual communities, it is customary to offer through mailing lists forums or Web sites supported by the community, third-party products which might be of interest to the group. In this way, the business supporting the communication infrastructure of the community is financed (e.g., the company which supports the CRM professional community *http://crm.insightexec.com/*). In addition, although this is a less frequent practice, some communities charge a fee in order to participate in debates (e.g., *http://www.well.com*).

A Source of New Clients

Membership to a community creates a favorable predisposition in a member toward others and toward the company supporting the platform for community interchange (Kardaras et al., 2003). For this reason, the community encourages members to recruit new members who ultimately will become new consumers.

Marketing Strategies and Managerial Recommendations

The exploitation of virtual communities may be implemented by means of two clearly different marketing strategies: offering support or becoming a member of the group.

Strategy 1: Offering Support to the Community

With this strategy, the company manages the platform on which the community exchanges take place. In this way, the company is not a member of the community but simply facilitates its existence by offering a Web site for communication purposes. For the development of this concept, the main step should be an analysis of the real possibilities of creating a community via the Internet. These possibilities derive either from the particular characteristics of a product or brand or else the pre-existence of the community off-line. In the first case, the company should define the aspects which characterize the product or brand to ascertain to what extent these are likely to create an attendant community (e.g., the company Apple, *http://www.apple.com/usergroups/*, exploits the characteristics of Apple computers). In the second case, a new better quality method of communication is offered to a group of individuals who are already linked outside the Internet (e.g., MarketingProfs.com, *http://www.marketingprofs.com*, facilitates communication between marketing consultants and educators).

Strategy 2: Become a Member of a Community

In this case, the company attempts to be perceived as a virtual community member. As in the first case, the initial step consists of determining whether or not the characteristics of the product or brand are likely to create a community. Nevertheless, the existence of a feasible virtual community is not sufficient to make the brand company just one more member of the group. In this respect, obtaining the level of membership of a community will depend on the trustworthiness perceived in the company. According to established literature on trustworthiness (e.g. Roy et al. 2001; Sako and Helper, 1997; Morgan and Hunt, 1994; ANderson and Narus, 1990; Dwyer, Schurr and Oh, 1987; Anderson and Weitz, 1989; Flavián & Guinalíu, 2003, 2003; Bhattacherjee, 2002; Cheung and Lee; 2001; Gefen, 2000; Kolsaker and Payne, 2002; Luo, 2002; Walczuch, Seelen and Lundgren, 2001; Geyskens, Steenkamp and Kumar, 1998; Sabel, 1993), the organization needs to demonstrate its honesty (i.e., that it is sincere and delivers its promises); benevolence (being concerned for the welfare of the other members of the community, not acting in an opportunistic manner, and attempting to have compatible objectives); and competence (that is, it has sufficient capability for its contributions to the community to be significant and to generate value). However, the success of the aforementioned strategies depends to a great extent on the monitoring of a series of managerial guidelines. Specifically, these actions might be determined in the light of the following issues:

- *Analyzing the members' needs*: Logically, the community should be created and managed according to the needs of its members, not the needs of the company which promotes it, the advertisers, or any other group not involved in the community (Cothrel & Williams, 1999).

- *Fostering self-management*: If self-management is technically impossible, they should try to create a situation in which the contents of the community are generated and published directly by its members, and this would entail greater commitment in the community's members.

- *Minimizing control*: In the same line as the previous point, the community should be able to grow freely, and it is not advisable to establish control mechanisms on how the community members should mix with each other or the topics they should discuss in their conversations (Cothrel & Williams, 1999). Some freedom in the contents of conversations and messages should be granted. Obviously, any content which may be offensive for the company or other members should be screened, but excessive control is not recommended. The community itself will create its own internal rules and will reject any participants who do not provide any value. In fact, it might be interesting to create a space where the members are free to talk about anything they want to, even though it has nothing to do with the community's primary purpose. It is also advisable to publish on the Web site the community's conduct rules as soon as they are defined.

- *Using the most suitable technological structure:* The type of technology and tools used to manage a virtual community logically has a lot of implications. In the literature, several authors (Guthrie, 2000; Whittaker, Issacs & O'Day, 1997) have recommended using technological systems which are flexible, easy to manage, and visually attractive.

- *Specializing roles:* Several sociological research studies have shown that inside a community it is common for individuals to adopt different roles which give the community a greater dynamism. Taking these findings as a reference, Cothrel and Williams (1999) have suggested the advisability of having this role assignment inside virtual communities. Some possible roles are (a) social weaver: individuals who introduce new members in the community; (b) moderator: a respected member who channels the debates in a suitable direction and regulates conversations; (c) knowledge manager: a member who evaluates and searches for useful resources for the community; (d) opinion leader: a respected member who defines the community's ideological tendencies (establishing a system of scoring the members' comments is an interesting possibility for the group to define who is a leader and who is not); (e) instigator: a member who, voluntarily and respectfully, proposes controversial discussion topics, thus encouraging participation.

Technological Tools for Virtual Communities Management

A virtual community may be managed with the help of various tools and technologies. We shall look at some of these now, together with suggestions as to how they might be used:

- *Discussion forums:* This tool enables members of a community to read messages sent to a Web site, choose a topic of interest (so that messages can be filtered), and reply to the messages that have been viewed. The way the users view messages will depend on the type of software that manages the database and posts the messages on the Internet. Thus, some programs post the messages chronologically while others group the messages by topic, creating trees that enable a debate to be followed. Finally, it should be pointed out that these systems need to be able to inform participants of the messages added the same day or in the past week or else send a summary of the messages via e-mail. Some systems can also inform an individual when somebody has answered a message posted previously.

- *E-mail groups:* Usually, virtual communities use e-mail groups or mailing lists, a fairly complex software system (list-bot), via which the messages posted by a community member are forwarded to all other members. Currently, the use of mailing lists as the only tool in virtual community management is fairly rare, although there are some collectives whose only source of contact is e-mail, for example, certain research groups (e.g., the ELMAR marketing researchers' list, *http://www.elmar-list.org/*). There are thousands of mailing lists, and they are often associated with the posting of newsletters. It is worth pointing out here that these lists should have some type of security mechanism to prevent (1) anyone who is not eligible from subscribing or (2) third parties from adding others to the list without their permission (one option here is to make use of a confirmation e-mail). Finally, it should be mentioned that these lists sometimes require a moderator to prevent spam from reaching the list.

- *Chat rooms*: A chat room is a tool that enables a group of individuals to converse via text messages in real-time. When talking of chat rooms, mention should be made of instant messaging (MSN Messenger and the like). These systems are an evolution of chat rooms, the main difference being that they afford greater privacy in conversations. Furthermore, improvements in the programming and the spread of broadband facilities mean that these systems are adding new functions, such as videoconferencing or the transmission of multimedia files. Instant messaging systems are experiencing a spectacular growth rate, and many companies are beginning to bring out improved solutions aimed at corporate environments.

- *MUD (Multiple User Dimension, Multiple User Dungeon, or Multiple User Dialogue)*: MUDs refer to applications that enable users to adopt the character they want and visit imaginary worlds where they take part with others in games or other types of activities (e.g., online games at *http://www.uo.com and http://secondlife.com/*).

- *Content management*: A content manager is a type of software which facilitates the management of a Web site, especially with the publication of the content on a site. The advantage of these programs is that they make it easy to manage not only the publication of the content but also other functions, such as e-commerce platforms or discussion forums. In addition, in some cases, there is the possibility of giving permission to users to do certain things, such as posting news, and for this reason, they are a useful tool for setting up self-managed communities (e.g., the Linux OS community, *http://usermodelinux.org/*). There are a great many programs on the market to manage Web site content (e.g., the systems produced by the well-known company Vignette, *http://www.vignette.com/*). Some of these are sold as standard packages, while others belong to Web site design companies and are not available to the general public. There is also the option of using free systems, such as Postnuke (*http://www.postnuke.com*). These open source systems are rapidly spreading in the development of communities and blogs (personal online diaries).

- *Peer-to-peer systems*: These systems (e.g., eMule or Bit Torrent) enable users to share large files. They usually include additional functions, such as instant messaging or chat facilities, and also have Web-based virtual communities. In these communities, one can find resources needed for making more effective use of the tool (e.g., *http://www.emule-project.net*).

- *Other technologies*: Technologies, such as the RSS format (Really Simple Syndication) or the SMS (Short Message Service), have been gaining ground recently as instruments of communication in virtual communities. RSS is a "format for syndicating news and the content of news-like sites, including major news sites like Wired, news-oriented community sites like Slashdot, and personal blogs" (Pilgrim, 2002, p. 000). SMS, on the other hand, is a service for sending short text messages to mobile phones.

Conclusions and Future Trends

In this chapter, we have analyzed the concept of the virtual community as a successful online business model. Various companies have based their activities on exploiting this type of social structure, obtaining in a large number of cases significant rates of success. Cases, such as that of eBay, show that the capacity of the Internet as a communications device for consumers is a characteristic likely to be economically exploited to a notable degree. Virtual communities have a strong influence on marketing in that, among other things, they facilitate the development of viral marketing strategies, serve as a source of strategic information, generate barriers to prevent the entry of new competitors, and increase consumer security and trust.

The development of virtual communities may be implemented by means of two distinct strategies. First, the business entrepreneur may offer the community a platform for virtual communication which increases the quality and frequency of group contact. A second strategy ensures that the company is not only the agent sustaining the communication infrastructure of the group but also a participant as one more members of the community. In this case, the company must act in a way that ensures that the rest of the community perceives it as a trustworthy entity so that group members are willing to become committed. Nevertheless, the success of these strategies is dependent on the monitoring of trends, such as helping self-management by minimizing control on the group dynamic and defining roles or using suitable technologies. With regard to the latter, this chapter has pointed out the main technologies for community management, such as e-mail, content managers, or the latest systems based on communication via the cell phone.

For the short-term future, research into the concept of virtual communities is faced with some interesting challenges. First, there is a need to analyze the influence that new methods of communication will have on how virtual communities evolve. Worthy of mention among these new methods of communication is the arrival of wireless Internet technologies (e.g., Wi-Fi, Bluetooth, UMTS, or i-Mode), as well as the recent spread of other devices (e.g., Tablet PC, PDA, or Smartphones). Wireless technology will afford community members new means of interaction, while at the same time opening the way to alternative lines of business. Furthermore, there is a need to examine in more detail the application of virtual community marketing strategies aimed at specific sectors. Particularly interesting here is the use of virtual communities in electioneering marketing strategies (Flavián & Guinalíu, 2004). Research has shown that through virtual communities, thousands of voters can be called on to attend a candidate's meeting in a city or to make financial donations. Virtual communities can also be used obtain up-to-date information of events, collect opinions on ideas proposed in the campaign, and recruit campaign volunteers, among other possibilities. Finally, it would be useful to analyze in more detail the role of virtual communities in corporate environments, knowledge management, or their integration into CRM systems.

Acknowledgment

The authors are grateful for the financial support received from the Aragon Regional Government, FUNDEAR, and the University of Zaragoza (UZ2002-SOC-06).

References

Alba, J., Lynch, J., Weitz, B., Janiszewski, O., Lutz, R., Sawyer, A., & Wood, S. (1997). Interactive home shopping: Consumer, retailer, and manufacturer incentives to participate in electronic marketplaces. *Journal of Marketing, 61*, 38-53.

Anderson, E., & Weitz, B. (1989). The use of pledges to build and sustain commitment in distribution channels. *Journal of Marketing Research, 29*, 18-34.

Anderson, J.C., & Narus, J.A. (1990). A model of distribution firm and manufacturer firm working partnerships. *Journal of Marketing, 54*, 42-58.

Barnatt, C. (2001). Virtual communities and financial services-online business potential and strategic choice. *International Journal of Bank Marketing, 16*(4), 161-169.

Bauer, H., Grether, M., & Leach M. (2002). Building customer relations over the Internet. *Industrial Marketing Management, 31*, 155-163.

Bhattacherjee, A. (2002). Individual trust in online firm: Scale development and initial test. *Journal of Management Information Systems, 19*(1), 211-241.

Cheung, C.M.K., & Lee, M.K.O. (2001). Trust in the Internet shopping: Instrument development and validation through classical and modern approaches. *Journal of Global Information Management, 9*(3), 23-35.

Clemente, J., & Escriba, C. (2003, May-June). Influencia del comercio electrónico en el sistema agroalimentario. *Distribución y Consumo*, 93-99.

Cohen, A.P. (1985). The symbolic construction of community. Chichester, UK: Ellis Horwood.

Cothrel, J., & Williams, R.L. (1999). Online communities: Helping them form and grow. *Journal of Knowledge Management, 3*(1), 54-60.

Dellarocas, C. (2001). *Building trust online: The design of reliable reputation reporting mechanisms for online trading communities* (Working Paper No. 101). MIT Sloan School of Management, Center for eBusiness. Retrieved October 26, 2004, from *http://ebusiness.mit.edu/research/papers/101%20Dellarocas,%20Trust%20Management.pdf*

Dwyer, F.R., Schurr, P.H., & Oh, S. (1987). Developing buying-seller relationships. *Journal of Marketing, 51,* 11-27.

Fernback, J., & Thompson, B. (1995). Virtual communities: Abort, retry, failure? Retrieved October 26, 2004, from *http://www.well.com/user/ hlr/texts/VCcivil.html*

Flavián, C., & Guinalíu, M. (2003). Antecedents and consequences associated with greater trust by users of a Website. *32nd European Marketing Academy Conference*, University of Strathclyde, Glasgow (United Kingdom).Retrieved March 20, 2004, from *http://perso.wanadoo.es/e/ GUINALIU/ficheros/EMAC_2003.pdf*

Flavián, C., & Guinalíu, M. (2004, May 20-23). The use of blogs: A productive strategic marketing tool for political parties. *Journal of Computer Mediated Communication.* Retrieved October 26, 2004, *from http:// perso.wanadoo.es/e/GUINALIU/ficheros/Blogs.pdf*

Gefen, D. (2000). E-commerce: The role of familiarity and trust. *OMEGA: The International Journal of Management Science, 28,* 725-737.

Geyskens, I., Steenkamp, J., & Kumar N. (1998). Generalizations about trust in marketing channel relationships using meta-analysis. *International Journal of Research in Marketing, 15,* 223-248.

Görsch, D. (2001). *Do hybrid retailers benefit from the coordination of electronic and physical channels?* Proceedings of the 9th European Conference on Information Systems, (Bled). Slovenia. Retrieved October 26, 2004, from *http://ecis2001.fov.uni-mb.si/doctoral/Students/ ECIS-DC_Goersch.pdf*

Guthrie, P. (2000, March). Creating communities online. *Computer Weekly,* 23 (March). Retrieved *http://www.computerweekly.com/ Article21645.htm*

Hagel, J., III, & Armstrong, A.G. (1997). *Net gain. Expanding markets through virtual communities.* Cambridge: Harvard Business School Press.

Hagel, J., III, & Singer, M. (1999). *Net worth. Shaping markets when customers make the rules.* Cambridge: Harvard Business School Press.

Hoffman, D.L., & Novak, T.P. (1996). *Marketing in hypermedia computer-mediated environments: Conceptual foundations* (Working Paper No. 1). Project 2000: Research Program on Marketing in Computer-Mediated Environments. Retrieved October 26, 2004, from *http://*

elab.vanderbilt.edu/research/papers/pdf/manuscripts/Conceptual
Foundations-pdf.pdf

Kardaras, D., Karakostas, B., & Papathanassiou, E. (2003). The potential of virtual communities in the insurance industry in the UK and Greece. *International Journal of Information Management, 23*, 41-53.

Kollock, P. (1999). The production of trust in online markets. Retrieved October 26, 2004, from *http://www.sscnet.ucla.edu/soc/faculty/kollock/papers/online_trust.htm*

Kolsaker, A., & Payne, C. (2002). Engendering trust in e-commerce: A study of gender-based concerns. *Marketing Intelligence and Planning, 20*(4), 206-214.

Luo, X. (2002). Trust production and privacy concerns on the Internet. A framework based on relationship marketing and social exchange theory. *Industrial Marketing Management, 31*, 111-118.

Morgan, R.M., & Hunt, S.D. (1994). The commitment: Trust theory of relationship marketing. *Journal of Marketing, 58*, 20-38.

Muñiz, A., & O'Guinn, T.C. (2001). Brand community. *Journal of Consumer Research, 27*(4), 412-432.

Pilgrim, M. (2002). What is RSS? Retrieved October 26, 2004, from *http://www.xml.com/pub/a/2002/12/18/dive-into-xml.html*

Rheingold, H. (1993). *The virtual community: Homesteading on the electronic frontier.* New York: Addison-Wesley.

Roy, M.C., Dewit, O., & Aubert, B.A. (2001). The impact of interface usability on trust in Web retailers. *Internet Research: Electronic Networking Applications and Policy, 11*(5), 388-398.

Sabel, C.F. (1993). Studied trust: Building new forms of cooperation in a volatile economy. *Human Relations, 46*(9), 1133-1170.

Sako, M., & Helper, S. (1997). Determinants of trust in supplier relations: Evidence from the automotive industry in Japan and the United States. *Journal of Economic Behaviour and Organization, 34*(3), 387-417.

Steinfield, C., Adelaar, T., & Lai, Y. (2002, January 7-10). Integrating brick and mortar locations with e-commerce: Understanding synergy opportunities. *Proceedings of the Hawaii International Conference on System Sciences (Big Island, Hawaii).* Retrieved March 20, 2004, *http://www.msu.edu/~steinfie/HICSS2002.pdf*

Walczuch, R., Seelen, J., & Lundgren, H. (2001). *Psychological determinants for consumer trust in e-retailing.* Proceedings of the 8th Research Symposium on Emerging Electronic Markets. Retrieved October 26, 2004, from *http://www-i5.informatik.rwth-achen.de/conf/rseem2001/papers/walczuch.pdf*

Wang, Y., Yu, Q., & Fesenmaier, D.R. (2002). Defining the virtual tourist community: Implications for tourism marketing. *Tourism Management, 23,* 407-417.

Webmergers. (2003). Internet companies three years after the height of the bubble. Retrieved October 26, 2004, from *http://www.webmergers.com/data/article.php?id=67*

Whittaker, S., Issacs, E., & O'Day, V. (1997). Widening the net. *SIGCHI Bulletin, 29*(3), 27-30.

Chapter XVI

An Online Consumer Purchase Decision Cycle

Penelope Markellou, University of Patras, Greece

Maria Rigou, University of Patras, Greece

Spiros Sirmakessis, Technological Educational Institution of Messolongi, Greece

Abstract

This chapter presents the overall consumer purchase decision cycle and investigates the issues that affect Web users from e-shop selection to product delivery and final assessment of the shopping experience. This process has been divided into three successive stages: outside the e-shop, inside the e-shop, and after sales. Each stage is analyzed on the basis of customer states and transition conditions, while special focus is set on abandonment factors. The chapter aims to provide a thorough insight to e-shop features that ensure customer satisfaction and those that may result in further enhancement of online shopping. The ultimate objective is to provide guidelines for designing successful e-shops and clarify success and failure factors.

Introduction

The Web is a powerful tool that has changed the way of conducting business providing companies and customers with limitless options and opportunities. Companies, in an effort to stay competitive in the new global economy, are increasingly expanding their activities to this new communication channel, which features as a factor of major profit potential. As a direct consequence of e-commerce, there is an emergence of a new consumer type, the online consumer or e-customer that uses the Internet for purchasing products and services (Solomon, 2001). Moreover, the online consumer is empowered with new exciting capabilities: searching globally for solutions (products or services), comparing available options, finding details and additional information, reading opinions of others that have already bought the product/service, and transacting online.

Ensuring e-customer satisfaction is a not a simple task. To a certain degree, e-customers behave online similarly to how they behave off-line, but in order to fully understand e-customer behavior, it is important to understand why people use the Internet for their purchases, the benefits and the drawbacks of online buying, and the identification of clusters of customers who share common attitudes, behavior, and preferences online (Blackwell, Miniard & Engel, 2000). According to Seybold and Marshak (1998) consumers prefer the Internet because it offers easier and faster shopping. Understanding the process of online decision making is important for developing e-business strategies and can provide guidance for deploying adequate marketing (Underhill, 2000). The traditional consumer purchase decision cycle has six stages (Windham & Orton, 2000): *stimulate* (realize the need), *consider* (collect ideas for potential solutions), *search* (choose category), *choose* (make selection), *buy* (make purchase transaction), and *buy again* (repurchase as needed). There also exist variations since in some cases, stages are collapsed or skipped.

Adapted to the Web context, this cycle can be merged to three stages: *confidence building*, where a consumer realizes that there is an alternative option for buying products or services, *skirmish*, when the consumer purchases for the first time, and *war*, when the consumer keeps on buying products or services (Zaltman, 2003).

Lee (2002) presented a behavioral model for the e-customer based on three distinct phases: building trust and confidence, online purchase experience, and after-purchase needs. The first phase examines issues connected to the Web site's brand name, authentication, reliability, credibility, privacy, and security. Intuitive navigation, searching facilities, product information, payment modes, usability, and convenience are consumer requirements that affect the second phase. The last phase relates to on-time delivery, customer support, technical support, availability of product warranty, and so forth. The combination of the three phases releases a behavioral model that may increase consumer trust and lead to more online purchases. Other cases have also been recorded in the international literature presenting parts of the behavior of the online customer (McEnally, 2002; Mowen & Minor, 2000; Windham & Orton, 2000). This chapter presents a perspective of the overall consumer purchase decision cycle. Additionally, it investigates the issues that affect Web users, from selecting a specific e-shop to the delivery of the product and the overall assessment of the shopping experience. This process has been divided into three successive stages: *outside the e-shop, inside the e-shop,* and *after sales*, with each stage analyzed on the basis of customer states and transition conditions. Special focus is set on identifying the potential abandonment factors.

Outside the E-Shop

The need for a specific product can be the first motivation for buying it, but enough motivation is required for choosing online shopping. Figure 1 models the process from the purchase stimulus to entering the e-shop. This relates to the consumer's general attitude and familiarity with online shopping, prior personal experiences, experiences of others, or brand familiarity (Lynch, 2000a).

Figure 1. From the purchase stimulus to entering an e-shop

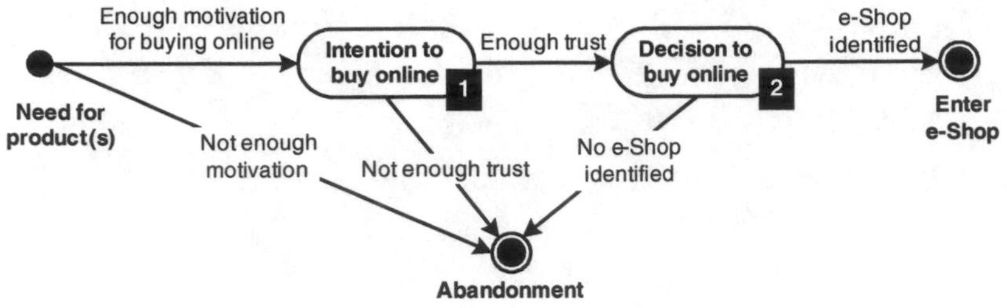

When the user decides to buy online, the next step is to identify an e-shop. New Web sites are launched everyday, and thousands of products are available online, so the competition for customer attention is fierce. Consumers are looking for a reliable and well known brand supporting the e-shop (Lynch, 2000b), are attracted by an aesthetically and consistent Web site design (Bouquet & Lynch, 2000), are enjoying navigating through a well-structured and usable Web site (Garrett, 2002; Sirmakessis, 2003), are convinced when receiving accurate content (Markellou et al., 2005a), and feel reassured by privacy statements and seals of secured transactions.

Additional factors that influence consumers when choosing an e-shop include pricing, quality of products and services, promotions, offers, after-sales support, and personalization facilities. There are also customers who chose an e-shop because a search engine returned this link high in its results list. A crucial issue for a company is to clearly identify its target group among online customers and give incentives to visit the Web site for the first time. Means for building awareness include affiliate programs with other Web sites, links from directory searches, e-mail notifications, banner advertisements, and so forth. (Markellou, Rigou & Sirmakessis, 2005b).

All the above mentioned apply mainly to new e-customers. For users with prior e-purchase experience, the issue of trust extends to the notion of loyalty. In this case, the Web site should try to capture relevant data that will allow it to personalize content, provide customized experiences, and support online communities. Without significant motivation and trust, the potential customer fails to identify an e-shop and abandons the process of online shopping before actually entering.

Inside the E-Shop

Figure 2 presents the steps e-consumers follow when navigating in an e-shop. It involves entering the e-shop, browsing/searching to identify the item(s) to place in the shopping cart, proceeding to the checkout to pay, and leaving the shop.

Registration/Login is an optional part of the process, but many sites use it for tracking individual consumers. First time visitors are asked to provide personal information (such as sex, age, income status, educational background, occupation, marital status, and other demographic information, as well as preferences, requirements, shipping address, billing details, etc.). The amount of information varies and depends on the kind of available products or services and the characteristics of the targeted consumer group. Login is the process a user has to go through in order to be identified by the system in each subsequent visit (after registration). Since Web site visitors are reluctant to reveal personal information and insecure about its use, e-shops should provide clear statements on their policies for the use of the collected information (disclosure policies) and the security precautions they have taken (authentication seals). Registration/Login may be postponed until before checkout so consumers are allowed to browse anonymously and identify themselves only when it is absolutely necessary.

The next step is to look for interesting items to be placed in the shopping cart either by browsing through products using the catalogue or by searching for something in particular. Web users are familiar with the notion of searching and the typical mechanism of browsing through search results. Regardless of how

Figure 2. The shopping process from entering to leaving the e-shop

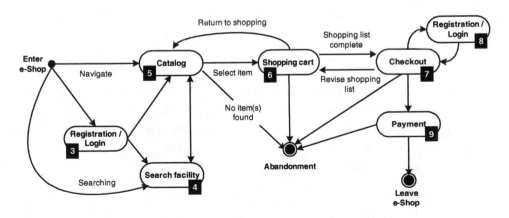

well structured and efficient the catalog is, an e-shop must also provide a search function and keep it available and visible on all pages so that customers may resort to it when needed. The general guideline is that the search should operate the way customers expect it to. It should be tolerant to minor misspellings, allow for synonyms and string keywords, and return an easily interpretable results page and links to product pages. Furthermore, in the case that a search failed, the response page should clearly indicate that no actual results are returned and, where possible, provide a way for customers to refine their search until they locate products (provided that the e-shop has them available).

On adding an item to the shopping cart, customers are usually transferred to the page with the current cart contents and price information. On that page, they should be allowed to remove one or more items, change item quantities, and automatically recalculate costs. In addition, a "return to shopping" link should exist to transfer them back to where they were before adding the last item to the cart (Markellou, Rigou & Sirmakessis, 2005c). This logic reflects exactly the real-life paradigm; while walking the corridors of a supermarket, we pick up items, put them in the cart and keep on walking until we find the rest of the items in our shopping list. Companies should construct an effective electronic shopping cart taking advantage of all conveniences offered by IT, stressing this comparative advantage over traditional shopping (Nielsen, 2000).

Checkout is the phase where they are asked to fill in information about delivery (mail address, date of delivery, packaging details, shipping options, etc.). If the user has not yet logged in (or is not obliged to login), the checkout phase asks for additional data that would otherwise be filled in automatically. Again, typing should be kept to a minimum, as in the previous cases, and all optional fields should be clearly and consistently marked. This also stands for certain information required during the payment phase, following right after checkout. Payment requires simplicity, clarity, certificates, reassurance, and a sense of professionalism and safety communicated in as many ways as possible. Security in transactions is not negotiable. There must be detailed information about the measures taken to guarantee security, and it must be available on every page of the payment process for whoever wants to view it. On successful completion of the payment phase, the customer should receive a confirmation through a response page or e-mail message that provides a code number (which uniquely identifies the submitted order) and the estimated delivery time. Using the order code number, the customer should be able to trace the status of the order at any time before receiving the product(s).

What Can Go Wrong?

A user may abandon a Web site for many reasons that do not necessarily have to do with the Web site. In this chapter, we deal only with reasons that are the e-shop's responsibility focusing on parameters that can be controlled and monitored.

- *Missing product or product details:* One of the most serious abandonment reasons is failure in locating the product. In such cases, an effective search utility or the assignment of products in multiple catalogue categories might solve the problem. If the product does not exist in the shop, the best approach is to inform customers as soon as possible. Missing or low quality product details is another major issue because it acts as an indication of poor professionalism and decreases the chances of making a sale.

- *Unreliable and inefficient e-shop operation:* The e-shop should be a 24/7 system, able to function under extreme traffic conditions, providing fast interaction and navigation. It should include simple and space efficient Web pages, representative and well-structured catalogue categories, and be available to users with low bandwidth connections.

- *Inconsistency:* The e-shop should be consistent in both the way it looks and functions. All pages should communicate the same business identity, and the way to interact and move around it must be consistent. For example, use the same buttons at the same position when they are meant to perform the same function.

- *Outdated technology, bad quality graphics, hard to read text:* It gives a good impression to users to see latest technological practices implemented. The effort, time, and money invested in setting up an e-shop are all indicated by the e-shop itself, and it is only natural to expect that a customer trusts the professionalism of a technologically advanced and carefully designed e-shop.

- *Complicated or ineffective search:* Search should be provided in combination with the catalogue in order to provide customers with a reliable alternative tool to locate products. Nielsen et al. (2000b) found that if users cannot find what they are looking for on their first search attempt, the odds that they will succeed in further searches decrease with each subsequent attempt. It is important for internal search engines to deliver relevant results on the first query.

- *Missing seals of approval, non-existent or insufficient details on security and disclosure policy, unclear company identification:* Security concerns have been the large obstacles for e-commerce. The e-shop should provide clear links to its identity, activities, background, address, telephone/fax numbers, and any other solid information regarding its physical status. Some e-shops even have photos of their personnel available, based on the principle that people like to do business with people. Seals, certificates, authentication signs, and detailed information on privacy should be included as well.

Ways to Further Enhance Online Shopping

For an e-shop to increase its chances of ensuring customer satisfaction and long-term profits, it must invest in advanced technological solutions, sense the needs of customers, and serve them efficiently. Typically, a personalized Web site recognizes its users, collects information about their preferences, and adapts its services in order to match user needs. In the e-commerce domain, personalization has spread widely and provides mechanisms to learn more about customer needs, identify future trends, and eventually increase customer loyalty. Personalization may be used in numerous ways for enhancing the online shopping experience. It can greet the user on revisiting, or, alternatively, it automatically prefills all data the user has already provided in registration. Using more sophisticated approaches (Markellou et al., 2005d), an e-shop can also provide:

- *Personalized product recommendations and marketing:* This refers to the recommendation of a set of products that are related to customer interests and preferences. Many e-shops today implement such techniques to cross-sell, up-sell, and suggest products a user might be interested in buying (Kobsa, Koenemann & Pohl, 2001; Mobasher, Cooley & Srivastava, 2000; Pierrakos et al., 2003). The algorithms underlying a product recommendation mechanism depend on the type of products a shop sells, the profiles of customers it transacts with, and the marketing policy.

• *Personalized pricing*: E-shops may offer different pricing and payment methods to different users, such as discounts or installments to loyal customers.

• *Personalized product configuration*: Personalization can be a powerful method of transforming a standard product into a specialized solution for an individual. In most of the cases, such features require customer input for customizing certain product parameters, but the customer can use the shop to specify exact requirements and receive a product configuration along with a price offer. However, consumers may need additional tools to answer their product inquiries fast and correctly and speed up their decision-making process.

• *Filtering tools:* These are used for narrowing a large item set down to a subset that satisfies a number of additional criteria. Nielsen et al. (2000a) refer to this category of tools as *winnowing tools*. The set of criteria that can be used by consumers for the filtering is predefined by e-shop designers and depends on the product distinguishing features, as well as the set of similar products the e-shop has to offer. Typical criteria comprise price ranges, brands, colors, sizes, and so forth.

• *Comparison tools:* Apart from filtering, there are cases of products that are difficult to compare even when the list of similar products along with their attributes is available and the purchase decision needs further assistance. These tools typically offer summarizing tables to easily compare products on a feature-by-feature basis (e.g., Nokia phones at nokia.com). It is important that the e-shop allows customers to specify the products to be compared or even the features of the products to base the comparison upon.

• *Customer community:* Another way to improve the impression a customer has about an e-shop stems from the use of the *customer community*, in the form of customer opinions. Returning customers' opinions are considered important for visitors or new customers. In many e-shops, such ratings appear on the product pages and not on the product category pages. It is much wiser though to provide visual rating indications on the product listing in category pages since customers can use them as a product comparison factor. Textual comments, on the other hand, due to space constraints, may well be restricted to product category pages.

Figure 3. Product receipt, customer support, and overall assessment

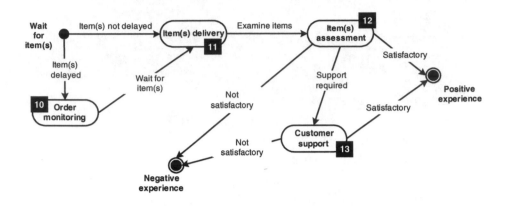

Aftersales

In the typical scenario, after successfully leaving the e-shop, customers enter the state of waiting to receive the goods (Figure 3). If the order is not delivered by the prespecified time, the customer usually invokes the order monitoring procedure. On receiving the order, the customer proceeds with assessing whether the products are in good condition and of satisfying quality. If the result of the assessment is negative, so is the impression of the online shopping experience and the specific e-shop. There are many cases of merchandise sold online (for example, computer hardware or software) that requires after-sales support. In such cases, the overall customer impression of the e-shop may be re-evaluated.

Conclusions

In this chapter, we examined the behavior of online customers assuming that it is triggered by a stimulus to purchase a certain item or set of items. Based on this assumption, we developed several models and drew a set of recommendations for all those whose decisions and objectives affect the online shopping experience. The figures presented, along with the investigation of the factors that determine the transition between states, also complied with this condition.

To enhance the online shopping experience and treat each customer individually, personalization has a central role. A big challenge, though, remains the lack of trust on the customer side and the lack of customer data on the e-shop side to base the personalization decisions on. Recording and predicting online consumer behavior is an arduous task. As Web surfers are increasingly extending their online experience, they become suspicious of any tracking/logging process. On the other hand, they become more demanding on special services focused on their individual needs. Research in the area of mining Web usage data (Sirmakessis, 2004) should be accompanied by security preservation methods to increase consumer confidence in the use of the Internet for selling and buying in the future.

References

Blackwell, R., Miniard, P.W., & Engel, J. (2000). *Consumer behavior*. Cincinnati, OH: South-Western College Publishers.

Bouquet, C., & Lynch, P. (2000a) Mood and Involvement Get Costomer Attention. *Online Consumer Behavior, 2*, Accenture Research Note. Available online at: *http://www.accenture.com*

Bouquet, C., & Lynch, P. (200b) Be Legit or Quit: Competing Locally on the Global Internet. Working Paper from Accenture Institute for Strategic Change.

Garrett, J. (2002). *The elements of user experience: User-centered design for the Web*. New York: American Institute of Graphic Arts in Conjunction with Indianapolis: New Riders.

Kobsa, A., Koenemann, J., & Pohl, W. (2001). Personalized hypermedia presentation techniques for improving online customer relationships. *Knowledge Engineering Review, 16*(2), 111-155.

Lee, P. (2002). Behavioral model of online purchasers in e-commerce environment. *Electronic Commerce Research, 2*(1/2), 75-85.

Lynch, P. (2000). How to win and influence buyers: Persuasion techniques that make Web sites sticky. *Online Consumer Behavior, 1*, Accenture Research Note. Available online at: *http://www.accenture.com*

Markellou, P., Markellos, K., Rigou, M., Sirmakessis, S., & Tsakalidis, A. (2005a). *E-business: From theory to action.* Athens, Greece: Ellinika Grammata.

Markellou, P., Rigou, M., & Sirmakessis, S. (2005b). Web personalization for e-marketing intelligence. In S. Krishnamurthy (ed.), *Contemporary research in e-marketing: Volume 1.* Hershey, PA: Idea Group.

Markellou, P., Rigou, M., & Sirmakessis, S. (2005c). Product catalog and shopping cart effective design. In Y. Gao (ed.), *Systems design and online consumer behavior.* Hershey, PA: Idea Group.

Markellou, P., Rigou, M., & Sirmakessis, S. (2005d). Mining for Web personalization. In A. Scime (ed.), *Web mining: Applications and techniques.* Hershey, PA: Idea Group.

McEnally, M. (2002). *Cases in consumer behavior: Volume 1.* Upper Saddle River, NJ: Prentice Hall.

Mobasher, B., Cooley, R., & Srivastava, J. (2000). Automatic personalization based on Web usage mining. *Communications of the ACM, 43*(8), 142-151.

Mowen, J.C., & Minor, M.S. (2000). *Consumer behavior: A framework.* Englewood Cliffs, NJ: Prentice Hall.

Nielsen, J. (2000). E-commerce user experience: Design guidelines for shopping carts, checkout, and registration (Unpublished report). Fremont, CA: Nielsen Norman Group.

Nielsen, J., Farrell, S., Snyder, C., & Molich, R. (2000a). E-commerce user experience: Product pages (Unpublished report). Fremont, CA: Nielsen Norman Group.

Nielsen, J., Molich, R., Snyder, C., & Farrell, S. (2000b). E-commerce user experience: Search (Unpublished report). Fremont, CA: Nielsen Norman Group.

Pierrakos, D., Paliouras, G., Papatheodorou, C., & Spyropoulos, C. (2003). Web usage mining as a tool for personalization: A survey. *User Modeling and User-Adapted Interaction, 13,* 311-372.

Pirolli, P., & Card, S. (1999). Information foraging. *Psychological Review, 106*(4), 643-675.

Seybold, P., & Marshak, R. (1998). *Customers.com: How to create a profitable business strategy for the Internet and beyond.* New York: Crown Business.

Sirmakessis, S. (2003). *Human computer interaction.* Athens, Greece: Ellinika Grammata.

Sirmakessis, S. (ed.). (2004). *Text mining and its applications. Studies in fuzziness and soft computing.* Berlin, Heidelberg, Germany: Springer-Verlag.

Solomon, M. (2001). *Consumer behavior: Buying, having, and being.* Upper Saddle River, NJ: Prentice Hall.

Underhill, P. (2000). *Why we buy: The science of shopping.* Australia: Simon & Schuster.

Windham, L., & Orton, K. (2000). *The soul of the new consumer: The attitudes, behaviours, and preferences of e-customers.* Oxford: Windsor.

Zaltman, G. (2003). *How customers think: Essential insights into the mind of the market.* Boston: Harvard Business School Press.

Glossary

Attitudes Toward Advertising: A predisposition to respond in a favorable or unfavorable manner to a particular advertising stimulus during a particular exposure occasion.

B2B (Business-to-business): The exchange of services, information, and/or products from one business to another, as opposed to between a business and a consumer.

B2C (Business-to-consumer): The exchange of services, information, and/or products from a business to a consumer, as opposed to between one business and another.

Business Model: A view of the business at any given point in time. The view can be from a process, data event, or resource perspective and can be the past, present, or future state of the business.

Computer Mediated Communication: A process of human communication via computers, involving people, situated in particular contexts, engaging in processes to shape media for a variety of purposes.

Culture: The knowledge and values shared by a society.

Customer Loyalty: The degree to which customers are predisposed to stay with your company and resist competitive offers.

Customer Relationship Management (CRM): Entails all aspects of interaction a company has with its customer, whether it is sales or service related. (Also referred to as **Customer Relationship Marketing**).

Customer Satisfaction: Measure or determination that a product or service meets a customer's expectations, considering requirements of both quality and service.

Customer Service: Activities and programs provided by the seller to make the relationship a satisfying one for the customer.

Customer Value: The difference between the benefits that a customer is receiving from the acquired products and services and the effort and cost that the customer has to invest to get the product. See **Lifetime Value (LTV).**

Distribution Channel The method through which a product is sold including retailers, cataloguers, Internet commerce Web sites, and so forth.

E-Commerce: Business that is conducted over the Internet using any of the applications that rely on the Internet, such as e-mail, instant messaging, shopping carts, Web services, UDDI, FTP, and EDI, among others. Electronic commerce can be between two businesses transmitting funds, goods, services, and/or data or between a business and a customer.

Electronic Marketing (E-Marketing): Using electronic means, such as the Internet, to market products and services.

Enterprise Resource Planning: A business management system that integrates all facets of the business, including planning, manufacturing, sales, and marketing. As the ERP methodology has become more popular, software applications have emerged to help business managers implement ERP in business activities, such as inventory control, order tracking, customer service, finance, and human resources.

E-Retailing (E-Tailing)?: Retailers selling products to customers over the Internet.

Experience Products: Those products for which either full information on dominant attributes cannot be attained without direct experience, or information search is more costly/difficult than direct product experience.

Forward Auctions: Auctions that tend to sell excess inventory at low prices and involve buyers bidding for inventory.

Internet: A global network connecting millions of computers.

Internet Marketing: See **Electronic Marketing.**

IT Infrastructure: The systems and network hardware and software that support applications. IT infrastructure includes servers, hubs, routers, switches, cabling, desktop, lap, and handheld devices.

Lifetime Value (LTV): The net profit a customer contributes to a business over the entire life cycle. Generally calculated as gross margin or contribution to overhead minus the promotional costs of acquisition and retention, including any discounts.

Marketing Strategy: The articulated plan for the best use of an organizations resources and tactics to meet its marketing objective.

Online Advertising: Advertising messages on the Internet.

Online Auction: A public sale online in which property (or an article) is sold to the highest bidder.

Online Marketing: See **Electronic Marketing**.

Online Shopping: Systems that provide a secure environment for browsing and purchasing products over an online service.

Reverse Auctions: Auctions that work with sellers bidding to arrange long-term contracts on standardized products.

Search Products: Products for which full information on the most important attributes can be obtained prior to purchase.

Supply Chain Management (SCM): The analysis of and effort to improve a company's processes for product and service design, purchasing, invoicing, inventory management, distribution, customer satisfaction and other elements of the supply chain. SCM usually refers to an effort to redesign supply chain processes in order to achieve streamlining.

Viral Marketing: The process of getting customers to pass along a company's marketing message to friends, family, and colleagues.

Virtual Community: A group sharing common interests in cyberspace rather than in physical space. Virtual communities exist in discussion groups, chat rooms, listservs, listprocs, and newsgroups.

World Wide Web (WWW): A system of Internet servers that supports specially formatted documents in a markup language called HTML (HyperText Markup Language).

About the Authors

Irvine Clarke III is an associate professor of marketing at James Madison University (USA) teaching international marketing and marketing management. He received his BSBA in Marketing from the University of Richmond and his MBA. and Ph.D from Old Dominion University. Prior to joining JMU, Dr. Clarke held the Freede Endowed Professorship of Teaching Excellence at Oklahoma City University. His current research in international marketing and marketing technology has been published in the *Journal of International Marketing*, *International Marketing Review*, *Industrial Marketing Management*, *Journal of Marketing Education*, *Marketing Education Review*, *Journal of Internet Commerce*, and the *Journal of the Academy of Marketing Science*. Dr. Clarke has taught at locations in Canada, England, France, Germany, Malaysia, Mexico, Singapore, and the People's Republic of China. He has 15 years of public and private sector organizational experience in various marketing areas. He currently serves on the editorial review boards of *International Marketing Review*, *Journal of Marketing Education*, *Health Marketing Quarterly*; also serving as section editor for *Marketing Education Review* and abstracts editor for *The Journal of Personal Selling & Sales Management*.

Theresa B. Flaherty is an associate professor of marketing at James Madison University (USA) where she teaches strategic internet marketing and integrated marketing communications. Prior to this position, she was an assistant professor of marketing and a member of the e-commerce faculty at Old Dominion University. She also taught marketing management and marketing research classes at the University of Kentucky where she earned her Ph.D. Dr. Flaherty, a certified e-marketing associate, has industry experience at JBI Customized Computer Solutions, California University of Pennsylvania's Entrepreneurial Assistance Center, IBM, and Service Corporation International. Dr. Flaherty has appeared in journals, such as *Industrial Marketing Management, International Marketing Review, Journal of Personal Selling and Sales Management, Journal of Internet Commerce, Marketing Education Review,* and *The Journal of Marketing Education.* Dr. Flaherty serves as a faculty advisor for Pi Sigma Epsilon, the national professional fraternity in marketing, personal selling, and sales management. Additionally, she serves as the Society for Marketing Advances Director of Placement Services, assistant editor for *MERLOT* (Multimedia Educational Resources for Learning and Online Teaching) and is the Web manager for *Marketing Education Review.*

* * * * *

Nicole Averill is vice president of marketing for Carmichael Productions (USA), a New York-based public relations, marketing, events, and promotions firm specializing in the film industry. She has both a BA in marketing and an MBA from Quinnipiac University.

Eileen Bridges is associate professor of marketing at Kent State University (USA), where she teaches service marketing, marketing management, and marketing research. She received her Ph.D in marketing from Northwestern University in 1987. Her research interests include e-commerce, services, and technology-based products, and she has published articles in the *Quarterly Journal of e-Commerce, Marketing Science,* the *International Journal of Research in Marketing,* the *Service Industries Journal, Psychology & Marketing,* and other journals. She also serves on the editorial boards of the *Service Industries Journal* and *Managing Service Quality.* Dr. Bridges is a member of INFORMS and the American Marketing Association, where she is currently an officer in the Services Special Interest Group.

Charles M. Brooks holds a Ph.D from Georgia State University and is associate dean of the School of Business and professor of Marketing at Quinnipiac University (USA). His research interests include electronic marketing, channel member relationships, retail site selection modeling, and geographic information systems. His work has been published in the *Journal of Consumer Behavior*, the *Journal of Retailing*, the *Journal of Business Research*, and the *Journal of the Academy of Marketing Science*, among others.

Michael K.M. Chiam is currently a senior lecturer with the School of Business and Accountancy in Ngee Ann Polytechnic. He has consulted with a number of organizations, both in the private and public sectors. The areas of his consulting include developing quality accreditation programs and strategic planning. Prior to joining the polytechnic, he has spent many years in the aeronautical and information technology industry. Michael is currently pursuing his marketing doctoral degree with the University of Western Australia (Australia). He holds an MS in quantitative business analysis from the Louisiana State University, USA (member of the Omega Rho, International Honor Society) and a BBA from the University of Oregon, USA.

Terry Daugherty (Ph.D, Michigan State University, USA) is an assistant professor in the Department of Advertising at the University of Texas at Austin. His research focuses on investigating consumer behavior and strategic media management, with work appearing in the *Journal of Advertising*, *Journal of Interactive Advertising*, *Journal of Consumer Psychology*, *Journal of Interactive Marketing*, and in the impending books *Online Consumer Psychology: Understanding and Influencing Consumer Behavior in the Virtual World*, and *Marketing Communication: Emerging Trends and Developments*. Prior to joining the Department of Advertising, he held a post-doctoral fellowship with eLab in the Owen Graduate School of Management at Vanderbilt University and has worked in advertising media as well as event marketing.

Matthew Eastin (Ph.D, Michigan State University) is an assistant professor in the School of Journalism and Communication at Ohio State University (USA). His research focuses on the social and psychological mechanisms that influence the uses and effects of media with a particular emphasis in exploring new media. Research topics include information processing in online environments, social and psychological antecedents of current Internet behavior and off-line outcomes, e-commerce, online gaming, and Internet literacy and children. His most recent work has appeared in *Computers in Human Behavior, CyberPsychology & Behavior, Journal of Broadcasting & Electronic Media, Journal of Computer Mediated Communication* and *Media Psychology*.

Carlos Flavián holds a Ph.D in business administration and is professor of marketing in the Faculty of Economics and Business Studies at the University of Zaragoza (Spain). His dissertation was awarded best thesis on marketing in Spain. His has been published in several academic journals, such as the *European Journal of Marketing, International Journal of Retail and Distribution Management, International Review of Retail, Distribution and Consumer Research, Journal of Retailing and Consumer Services, Journal of Consumer Marketing* or *Journal of Strategic Marketing*, and different books, such as *The Current State of Business Disciplines, Building Society Through e-Commerce,* and *Contemporary Problems of International Economy*.

Harsha Gangadharbatla is a doctoral candidate in the Department of Advertising at the University of Texas at Austin (USA). His research focuses on investigating new and alternative media and has appeared at numerous marketing and communication conferences, including the American Academy of Advertising and American Marketing Association, among others. More recently, his research has been published in the *Journal of Interactive Advertising*.

Ronald E. Goldsmith received his Ph.D from the University of Alabama at Tuscaloosa and is the Richard M. Baker professor of marketing at Florida State University (USA). His interests include consumer behavior, marketing research, services marketing, and strategic marketing. He also consults with businesses and government agencies. His publications include approximately 150 journal articles and conference papers as well as a book, *Consumer Psychology for Marketing*, which is currently in its second edition. His articles have appeared in the *Journal of the Academy of Marketing Science*, the *Journal of Business Research*, the *Service Industries Journal*, and others. He is currently the North American editor of the *Service Industries Journal*.

Miguel Guinalíu is assistant professor in the Faculty of Economics and Business Studies (University of Zaragoza, Spain). Previously, he worked as an e-business consultant. His main research line is online consumer behavior, particularly the analysis of online consumer trust and virtual communities. His work has been presented in national and international conferences and has been published in several journals and books.

Jari H. Helenius is a key account manager of DNA Finland Ltd., part of the Finnet Group of Companies and a doctoral candidate in the Swedish School of Economics and Business Administration (Hanken) in Helsinki (Finland). He received his Msc. Economics degree from Hanken in 2002, based on a thesis entitled "Developing Brand Equity through Wireless Devices -Case Coca-Cola's 'Red Collection' Campaign." His main research interests include marketing enabled by new technology, branding, and private label services. He has lived and studied in the U.S. and was an exchange student at Western Washington University, Bellingham.

Charles F. Hofacker is a professor of marketing at Florida State University (USA), where he teaches courses in e-commerce, supply chain marketing, quantitative methods, and operations research. He received his Ph.D in mathematical psychology from UCLA in 1982, and his current research interests include electronic marketing, especially consumer behavior in online environments. He has published in such journals as the *Journal of Marketing Research*, *Management Science*, the *Journal of Interactive Marketing*, the *Journal of the Academy of Marketing Science*, and *Psychometrika*. Dr. Hofacker belongs to the American Marketing Association, the Academy of Marketing Science, INFORMS, the European Marketing Academy, and the Association for Computing Machinery.

Gopalkrishnan Iyer is an associate professor of marketing and the director for the Center for Services Marketing at the Florida Atlantic University (FAU) in Boca Raton, Florida (USA). After an engineering undergraduate degree and an MBA, he received his Ph.D in marketing from Virginia Tech in 1993. He has taught at Virginia Tech, Baruch College (City University of New York), ESIC (Spain), and Dartmouth College. His industry experience includes strategic management positions for two of Asia's top 50 multinational corporations. His current research interests are in areas of international business, business-to-business marketing, e-business, and the Hollywood film industry.

Mark R. Leipnik (Ph.D, University of California at Santa Barbara) is an associate professor of geography at Sam Houston State University (USA) and director of the GIS research lab for TRIES. He has authored a textbook, six book chapters, and over 50 articles in the areas of application of GIS to criminal justice, marketing, and the use of GIS and related technologies in environmental assessment. Dr. Leipnik has worked as a GIS environmental consultant and/or analyst for local, county, state, and federal governments and for large and small businesses and consulting firms. He holds an MBA from Rice University and an undergraduate degree from the University of California at Santa Barbara.

Veronica Liljander is professor of marketing and head of the Department of Marketing and Corporate Geography at the Swedish School of Economics and Business Administration (Hanken) in Helsinki (Finland). She earned her Ph.D from Hanken in 1995, and her current research interests include service and relationship marketing, consumer responses to electronic services, and private label marketing. She has published articles, for example, in the *International Journal of Service Industry Management, Journal of Services Marketing,* and *Psychology & Marketing.* In 2000-2001, she spent a year at Maastricht University in The Netherlands, and in 2004, two months at the University of Roskilde in Denmark.

Adam Lindgreen finished an MSc at the Technical University of Denmark after graduating in chemistry, engineering, and physics. He completed an MBA at the University of Leicester and a Ph.D in marketing at Cranfield University with 18 months as a visiting research fellow at the University of Auckland. Dr. Lindgreen is now with Eindhoven University of Technology (The Netherlands). He has consulted for a number of organizations, has taught at numerous universities, and is a visiting professor/reader with Groupe HEC. He has published in journals including *Psychology and Marketing, Journal of Marketing Management, Journal of Customer Behaviour, Journal of Brand Management,* and *Journal on Relationship Marketing.*

Michael W. Little, associate professor of marketing and former associate dean of graduate studies in business, teaches retailing and marketing strategy. He has consulted and provided training with area retailers, banks, and small businesses. He is a member of the American Marketing Association, American Collegiate Retailing Association and Richmond, Virginia International Rotary

Club. Dr. Little has published in the American Marketing Association Proceedings, Academy of Marketing Science Proceedings, *Journal of Business Research, Journal of Retailing, Journal of Services Marketing, Journal of Health Care Marketing,* and numerous other publications. His current area of research interest is in electronic retail/marketing on the World Wide Web.

Angelo Manaresi is a full professor of marketing and management, Department of Management, Faculty of Economics, University of Bologna (Italy). Dr. Manaresi graduated with honors in economics at the University of Bologna and obtained a Ph.D in economics and marketing from the London Business School in 1993. His research interests include the management of distribution networks at the international level, consumer behavior and hedonic consumption, and loyalty and branding.

Chris Manolis, Ph.D is an associate professor of marketing at Xavier University (USA). Dr. Manolis received a BA in psychology from the University of California at Los Angeles, an MBA from San Francisco State University, and a Ph.D from the University of Kentucky. He has taught marketing courses both at the graduate and undergraduate levels, and his research interests include the study of psychological and behavioral processes of exchange participants along with various methodological/empirical research issues. Dr. Manolis' research has appeared in journals, such as the *Journal of the Academy of Marketing Science, Journal of Business Research*, and *Journal of Consumer Psychology,* and *Applied Social Psychology.*

Penelope Markellou has an MSc in computer science in the area of designing and evaluating e-business systems. She is working as a researcher in the Computer Engineering and Informatics Department of the University of Patras (Greece) and in the Internet and Multimedia Technologies Research Unit of the Research Academic Computer Technology Institute. Her research interests lie in personalization and Web mining techniques applied in the e-commerce and e-learning domains. She has published several research papers and is coauthor of five book chapters.

Sanjay S. Mehta (MS, Ph.D, University of North Texas) is an associate professor of marketing and recipient of the 2004 Excellence in Research Award at Sam Houston State University (USA). He has presented and/or published over 120 articles in the areas of franchising, global marketing, e-commerce/e-marketing, sports marketing, and GIS. His work has been published in numerous journals including: *Journal of Asian Business, The Cornell H.R.A. Quarterly, Journal of Professional Services Marketing, Health Marketing Quarterly, Journal of Customers Services in Marketing and Management, Journal of Business Strategies*, and *Journal of Business and Entrepreneurship*. Dr. Mehta holds a BS in mathematics and an MBA in management from Angelo State University.

Chris O'Leary is a project manager at MSI Business Systems Pty Ltd, Sydney (Australia). Dr. O'Leary has been involved in many management consulting projects in the United States, the United Kingdom, and Australia. His research interests include data management, Internet marketing, and information system management.

Maria Rigou has an MSc in computer science in the field of interactive systems evaluation. She is currently working as a researcher (Ph.D student) at the Computer Engineering and Informatics Department of the University of Patras (Greece) and also at the Internet and Multimedia Technologies Research Unit of the Research Academic Computer Technology Institute. As a researcher, she is working on personalization techniques and the use of Web usage mining in adapting Web site content and structure and has several publications in the domain of Web mining and its applications in e-commerce, e-learning, as well as the formation and behavior of online communities. She is coauthor in three book chapters on personalization related technologies.

Sally Rao is a lecturer in marketing at School of Commerce at Adelaide University (Australia). Dr. Rao is an active researcher and is involved in both commercial and academic research in the areas of relationship marketing, Internet marketing, consumer behavior, and services marketing. She has published in international journals, such as the *European Journal of Marketing, the Journal of Business and Industrial Marketing, Australasian Marketing Journal,* and *Qualitative Market Research: an International Journal.*

Francesca Dall'Olmo Riley earned a BA from University of Bologna, a PG Dip. from the London School of Economics, and Ph.D from the London

Business School. Dr. Riley worked in marketing management and direct marketing for several years in Italy, the USA, and the UK before obtaining a Ph.D in consumer behavior from London Business School. She is now a reader in marketing at Kingston University Business School (UK). Her research interests include attitudes, branding, online shopping, and customer relationship management issues. She has published widely in international journals.

Mary Lou Roberts is professor of marketing at the University of Massachusetts Boston. Her Ph.D in marketing is from the University of Michigan. Dr. Roberts recently published *Internet Marketing: Integrating Online and Offline Strategies* (McGraw-Hill, 2003). She is senior author of *Direct Marketing Management, second edition* (Prentice-Hall, 1999). She has published extensively on a variety of marketing topics and currently serves on a number of editorial review boards of U.S. and European journals. Her current research deals with information-driven marketing strategies. She is a frequent presenter on programs of both professional and academic marketing organizations and has consulted for a wide variety of corporations and nonprofit organizations.

Jari Salo received his MSc (economics and business administration) degree from the University of Oulu (Finland) in 2002 and is currently finalizing his doctoral studies at the Finnish Graduate School of Marketing (FINNMARK). He has previously published in *E-Business Review* and in academic conferences including: HICSS, ANZMAC, EMAC, and IMP. His present research interests include digitization of business relationships and networks, electronic commerce, and mobile advertising.

Daniele Scarpi completed classical studies and graduated in economics at the University of Bologna (Italy), where he also earned a Ph.D in business economics in 2004. His research interests and publications focus on hedonic behavior, distribution channels, Internet shopping, and presentation order effects. Currently, he is a research fellow at the University of Bologna, where he also teaches service marketing.

Spiros Sirmakessis is an assistant professor in the Technological Educational Institution of Messolongi and the manager of the Internet and Multimedia Technologies Research Unit of the Research Academic Computer Technology

Institute (http://www.cti.gr). He is also an adjunct assistant professor in the University of Patras, teaching human computer interaction and e-business courses. His research areas include human computer interaction, human factors, e-business strategies, efficient techniques for designing interactive systems, and Web mining. He is coauthor of three books and several research papers published in international journals and conferences.

Jaana Tähtinen is acting professor at the University of Oulu (Finland), where she also received her Ph.D in 2001. Her thesis deals with the dissolution process of a business relationship in the tailored software business. Before joining the faculty, she worked within the industry and service sector as well as a director in a local science center. She has published in the *European Journal of Marketing, Industrial Marketing Management, International Journal of Service Industry Management, Journal of Market Focused Management*, and *Marketing Theory*.

Joëlle Vanhamme earned MA degrees in management and psychology, as well as a Ph.D in marketing from the Catholic University of Louvain. She also holds a CEMS master. She has been awarded two best paper awards, as well as a prize for her Ph.D thesis. Joëlle has worked as a researcher, lecturer, and consultant in numerous countries. She is now an assistant professor with the Erasmus University Rotterdam (The Netherlands) and a member of the ERIM Research Institute. Joëlle has published, among others, *in Psychology and Marketing, Journal of Marketing Management, Journal of Economic Psychology*, and *Recherche et Applications en Marketing*.

Heiko de B. Wijnholds is a former chair of the Marketing Department at Virginia Commonwealth University. He has authored and coauthored numerous publications in professional journals, conference proceedings, and textbooks. Among others, he has published in such journals as the *Journal of International Business*, the *Journal of Services Marketing*, the *Journal of Euromarketing*, and the *Journal of Professional Services Marketing*. Dr. Wijnholds has also presented (published) papers at annual and special meetings of leading academic and professional organizations, such as the American Marketing Association, the Academy of Marketing Science, the Academy of International Business and the European Society for Opinion and Marketing Research.

Alvin Y.C. Yeo conceptualizes and drives national programs to catalyze the e-business and security industry in his current role as a business development manager with the Infocomm Development Authority of Singapore. Prior to this, he assumed management responsibilities at Overseas Union Bank in branch operations, commercial banking, and segment marketing. Alvin is currently pursuing his marketing doctoral degree with the University of Western Australia (Australia). A former recipient of two ASEAN scholarships awarded by Singapore's Ministry of Education, he holds a master of business administration (distinction) from the University of Leicester and a bachelor of business administration in finance (merit) from the National University of Singapore.

Michael T. Zugelder joined Old Dominion University's College of Business and Public Administration (USA) in 1995 and currently serves as its business law and legal environment professor. His teaching includes courses in business law, electronic commerce, and employment law, and regulation at the undergraduate and graduate levels. His research and publications have focused on intellectual property and contract law, with particular regard to the Internet. Zugelder received his doctor of jurisprudence (JD), *cum laude*, from the University of Toledo in 1980 and his BA and MBA degrees from Indiana University in 1974 and 1976.

Index